BASIC SKILLS WITH DECIMALS AND PERCENTS

A Step-by-Step Approach

JERRY HOWETT

CAMBRIDGE Adult Education
PRENTICE HALL CAREER & TECHNOLOGY
Englewood Cliffs, New Jersey 07632

Prentice-Hall, Inc.
A Simon & Schuster Company
Englewood Cliffs, New Jersey 07632

Printed in the United States of America

10 9

ISBN 0-8428-2118-X

Prentice-Hall International,(UK) Limited, *London*
Prentice-Hall of Australia Pty. Limited, *Sydney*
Prentice-Hall Canada Inc., *Toronto*
Prentice-Hall Hispanoamericana, S.A., *Mexico*
Prentice-Hall of India Private Limited, *New Delhi*
Prentice-Hall of Japan, Inc., *Tokyo*
Simon & Schuster Asia Pte. Ltd., *Singapore*
Editora Prentice-Hall do Brasil, Ltda., *Rio de Janeiro*

Contents

Introduction

In this book you will gain skill in adding, subtracting, multiplying, and dividing decimals and in using percents.

Here's how the book will help you. First, work the problems called "Step One to Decimal Skill" on page 2. This will tell you what parts of the decimal section of the book to study. When you finish the decimal section, work the problems called "Step One to Percent Skill" on page 52. This will tell you what parts of the percent section of the book to study. In both sections there are plenty of practice problems to sharpen your skills. You should work all the problems and check your answers. Correct answers are given at the back of the book.

There is truth in the old saying, "Practice makes perfect." The more you practice, the stronger your skills become. At the end of each section there is a review. These reviews will show you how much the book has helped.

All through the book there are word problems. They will give you a chance to use your skills in real-life problems. Some of the word problems give you practice in using charts and graphs like those in newspapers and magazines. The pages called "More Application Problems" ask you to use more than one skill. For these problems the answer section will show you step-by-step solutions.

Take your time as you work with this book. The strong skills you build now will make other kinds of math easier.

Step One to Decimal Skill

The problems on pages 2 through 6 will help you find out which parts of the decimal section of this book you need to work on. Do all the problems you can. At the end of the problems look at the chart on page 7 to see which page you should go to next.

1. Circle the mixed decimals in this list.

76.3 .479 8.067 92 1.2804

2. For each decimal or mixed decimal write how many decimal places the number has.

a. 2.977 ________ .0078 ________ 163.5 ________

b. 13.21 ________ 19.3 ________ 1.0742 ________

3. Cross out the zeros that are not needed in each decimal or mixed decimal.

09.0070 .00380 014.0601 080.05030

4. Write these decimals or mixed decimals in words.

a. .4 ________ 1.27 ________

b. .18 ________ 2.015 ________

c. .0004 ________ 16.09 ________

d. .000012 ________ 120.0056 ________

5. Write each number as a decimal or a mixed decimal.

a. sixteen hundredths ________

b. ten and five tenths ________

c. twenty-two hundred-thousandths ________

d. three hundred nine and eight thousandths ________

e. four hundred eight millionths ________

6. Circle the bigger decimal in each pair.

a. .081 or .8 | .05 or .061 | .203 or .03

b. .2 or .198 | .41 or .149 | .08 or .092

7. Circle the biggest decimal in each group.

a. .031, .3, or .03 | .8, .808, or .07 | .101, .01, or .11

b. .71, .072, or .7 | .06, .056, or .05 | .38, .4, or .401

8. Change each decimal or mixed decimal to a fraction or mixed number. Reduce each fraction.

a. .08 = | .3 = | 4.2 = | 9.125 =

b. .036 = | 8.0045 = | .006 = | 12.4 =

9. Change each fraction to a decimal.

a. $\frac{3}{5} =$ | $\frac{7}{10} =$ | $\frac{6}{25} =$ | $\frac{2}{9} =$

b. $\frac{3}{20} =$ | $\frac{3}{8} =$ | $\frac{1}{15} =$ | $\frac{5}{12} =$

10. .7 + .2 + .6 =

11. .3 + .24 + .812 =

12. .52 + .9 + .008 =

13. .305 + .6 + .0471 =

14. 8.3 + .556 + 4 =

15. .79 + 25 + 3.8 =

16. .029 + 15 + .36 =

17. 2.4 + 1.76 + .083 =

18. In 1970 there were 16.3 million people living in the New York City area. Experts believe in 1990 there will be 3.7 million more people living there. How many people will be living in New York City in 1990?

19. On Monday Carl bought 8.2 gallons of gas for his car. On Thursday he bought 7 gallons. On Sunday he bought 8.3 gallons. How many gallons of gas did Carl buy altogether that week?

20. .4 − .38 =

21. 1.2 − .64 =

22. 9.2 − 1.375 =

23. 6 − .28 =

24. 14 − .709 =

25. 3.6 − 1.44 =

26. An empty box weighs .025 kilograms. When the box is filled with cans of soup it weighs 4.62 kilograms. What is the weight of the cans inside the box?

27. In 1950 the world population was about 2.5 billion people. In 1980 the world population was about 4.47 billion. By how much did the world population increase from 1950 to 1980?

28. $8 \times .457 =$

29. $3.2 \times 4.9 =$

30. $.67 \times 2.5 =$

31. $.008 \times 3.6 =$

32. $.0045 \times 26 =$

33. $.348 \times 9.3 =$

34. $10 \times .32 =$

35. $6.7 \times 100 =$

36. $1000 \times 3.45 =$

37. One pound of round roast costs $2.20. How much does Verva have to pay for 2.6 pounds of round roast?

38. Paul can drive an average of 19.6 miles on one gallon of gas. How many miles can he drive on 8.4 gallons of gas?

39. $4.77 \div 9 =$

40. $361.2 \div 42 =$

41. $2.52 \div .6 =$

42. $2.45 \div .07 =$

43. $7.32 \div .012 =$

44. $.552 \div 2.4 =$

45. $.152 \div 1.9 =$

46. $45.6 \div .38 =$

47. $14 \div .35 =$

48. $72 \div 1.6 =$

49. $.48 \div 10 =$

50. $6.2 \div 100 =$

51. $4.357 \div 1000 =$

52. Fred bought 6.8 meters of lumber for $25.84. What was the price of one meter of lumber?

53. There are 2.2 pounds in a kilogram. Frank weighs 187 pounds. What is his weight in kilograms?

Check your answers on page 100. Then complete the chart below.

Problem numbers	Number of problems in this section	Number of problems you got right in this section	
1 to 9	9	______	If you had fewer than 7 problems right, go to page 8.
10 to 19	10	______	If you had fewer than 8 problems right, go to page 16.
20 to 27	8	______	If you had fewer than 6 problems right, go to page 20.
28 to 38	11	______	If you had fewer than 8 problems right, go to page 25.
39 to 53	15	______	If you had fewer than 12 problems right, go to page 33.

What Is a Decimal?

Decimals are a kind of fraction. You have used decimals since you first handled money. \$.60 is a decimal. It is 60 of the 100 equal parts in a dollar. When we write \$.60 like a common fraction, it is $\frac{60}{100}$. The denominator 100 tells us how many parts are in one dollar. The numerator 60 tells us how many parts we have.

Decimals are different from common fractions in two ways. One difference is that the denominators of decimals are not written. The other difference is that only some numbers — 10, 100, 1000, etc. — can be decimal denominators. Decimal denominators get their names from the number of **places** at the right of the decimal point. A place is the position of a digit. The decimal point itself does not take up a decimal place. .60 is a decimal with two places. .3942 is a decimal with four places.

The list below gives the names of the first six decimal places and the number of places they use. Memorize this list before you go on.

Decimal	Number of places
tenths	one place
hundredths	two places
thousandths	three places
ten-thousandths	four places
hundred-thousandths	five places
millionths	six places

The next list gives examples of decimals.

Decimal		Common Fraction
.8	eight tenths	$\frac{8}{10}$
.07	seven hundredths	$\frac{7}{100}$
.004	four thousandths	$\frac{4}{1000}$
.0001	one ten-thousandth	$\frac{1}{10{,}000}$
.00003	three hundred-thousandths	$\frac{3}{100{,}000}$
.000009	nine millionths	$\frac{9}{1{,}000{,}000}$

Mixed decimals are numbers with digits on both sides of the decimal point. $9.49 is a mixed decimal. It means 9 whole dollars and $\frac{49}{100}$ of a dollar. Mixed decimals have whole numbers at the left of the decimal point.

1. In this list circle the mixed numbers. Answers are on page 101.

.282	904.6	7.3	.729	.1037
40.9	.718	6.622	86.33	3.8

2. For each decimal or mixed decimal write how many decimal places the number has. Remember that decimal places are digits at the **right** of the decimal point.

6.75 ________	23.158 ________	.12406 ________
.51 ________	467.7 ________	3.29 ________
864.2 ________	2.68 ________	10.0854 ________
.039 ________	1.0899 ________	14 ________

Zeros are sometimes hard to understand in decimals. Any zero at the right of a decimal point and, at the same time, at the left of another digit is important. For example, all the zeros in this decimal are important: .00306; these zeros keep 3 in the thousandths place and 6 in the hundred-thousandths place. But in the decimal .04500, the two zeros at the right are useless. They have nothing to do with the value of the 4 and 5. We could rewrite this decimal as .045.

3. In this list cross out the zeros that are not needed in each decimal or mixed decimal.

90.0850	096.30	05.90	200.040
4.30600	30.0105	2,060.70	750.800
02.3580	6.0	.007050	60.2000

Reading Decimals

Remember that a decimal gets it name from the number of places at the **right** of the decimal point. To read a decimal, count the places at the right of the point.

EXAMPLE: Read the decimal .063

Step 1. Notice that every digit is at the right of the decimal point. There is no whole number with this decimal.

Step 2. Count the decimal places. The decimal has three places. Three decimal places are thousandths.

Step 3. Read .063 as **sixty-three thousandths**.

With mixed decimals, separate the whole number and the decimal with the word **and**.

EXAMPLE: Read 14.08

Step 1. Notice that there are two digits on each side of the decimal point. This number is a mixed decimal.

Step 2. Read the whole number as **fourteen**.

Step 3. Count the decimal places. The decimal part has two places. Two decimal places are hundredths.

Step 4. Read 14.08 as **fourteen and eight hundredths**.

Write these decimals or mixed decimals in words. Answers are on page 101.

1. .2 ______________________ .6 ______________________

2. .03 ______________________ .49 ______________________

3. .018 ______________________ .207 ______________________

4. .006 ______________________ .483 ______________________

5. .0058 ______________________ .0132 ______________________

6. .00325 ______________________ .000005 ______________________

7. .00009 ______________________ 1.6 ______________________

8. 208.4 ______________________ 13.09 ______________________

Writing Decimals

When you write decimals, decide how many places you need. Use zeros in places that are not filled.

EXAMPLE: Write nine thousandths as a decimal.

Step 1. Decide how many places you need. Thousandths need three places.

.009 *Step 2.* The digit 9 needs only one place. Use zeros in the first two places.

EXAMPLE: Write thirty-eight hundred-thousandths as a decimal.

Step 1. Decide how many places you need. Hundred-thousandths need five places.

.00038 *Step 2.* The number 38 needs only two places. Use zeros in the first three places.

When you write mixed decimals, put a decimal point in place of the word **and**.

EXAMPLE: Write nineteen and four hundredths as a mixed decimal.

Step 1. Write the whole number 19.

Step 2. Decide how many decimal places you need. Hundredths need two places.

Step 3. The digit 4 needs only one place. Use a zero in the first decimal place.

19.04 *Step 4.* Separate the whole number and the decimal with a decimal point.

Write each number as a decimal or mixed decimal. Answers are on page 101.

1. two tenths ____________ seven and five hundredths ____________

2. fifteen thousandths ____________ forty and three tenths ____________

3. six hundredths ____________ thirteen and seven thousandths ____________

4. nineteen ten-thousandths ____________ nine millionths ____________

5. two and eight hundred nine millionths ____________

6. five hundred twelve and four tenths ____________

7. eight hundred and six ten-thousandths ____________

8. eighteen hundred-thousandths ____________

Comparing Decimals

When you compare fractions, first change the fractions to new fractions with a common denominator. When you compare decimals, first change the decimals to new decimals with the same number of places. Decimals with the same number of places have a common denominator. You can put zeros to the right of a decimal without changing its value. .5 and .50 have the same value. The 5 is in the tenths place in each decimal.

EXAMPLE: Which decimal is bigger: .09 or .3?

Step 1. Put a zero at the right of .3 to change it to .30. Both decimals are hundredths now.

Step 2. Decide which is bigger, .09 or .30. Thirty hundredths is bigger than nine hundredths. .3 is bigger of the original two decimals.

EXAMPLE: Which decimal is biggest: .45, .4, or .405?

Step 1. Put a zero at the right of .45 to change it to .450.

Step 2. Put two zeros at the right of .4 to change it to .400. All three decimals are thousandths now.

Step 3. Decide which is biggest: .450, .400, or .405. Four hundred fifty thousandths is the biggest. .45 is the biggest of the original decimals.

Circle the bigger decimal in each pair. Answers are on page 101.

1.	.7 or .75	.16 or .2	.08 or .063
2.	.186 or .3	.005 or .05	.06 or .072
3.	.53 or .515	.22 or .212	.453 or .44
4.	.3 or .0458	.61 or .169	.09 or .1024

Circle the biggest decimal in each group.

5.	.4, .04, or .34	.205, .21, or .2	.0034, .403, or .34
6.	.18, .2, or .201	.57, .075, or .507	.011, .01, or .1101
7.	.07, .017, or .71	.6, .066, or .606	.404, .44, or .04
8.	.52, .5205, or .5	.08, .708, or .8	.3, .03, or .003

Changing Decimals to Fractions

A decimal is a kind of fraction. To change a decimal to a common fraction, write the digits in the decimal as the numerator. Write the name of the decimal (tenths, hundredths, thousandths, etc.) as the denominator. Then reduce the fraction.

EXAMPLE: Change .08 to a fraction.

$\underline{08}$	*Step 1.*	Write 08 as the numerator.
$\frac{08}{100}$	*Step 2.*	.08 has two decimal places. Two decimal places are hundredths. Write 100 as the denominator.
$\frac{08}{100} = \frac{2}{25}$	*Step 3.*	Reduce $\frac{08}{100}$ by 4. You do not need to write the zero in the numerator.

EXAMPLE: Change 7.6 to a mixed number.

$7\,\underline{6}$	*Step 1.*	Write 7 as a whole number, and write 6 as the numerator.
$7\frac{6}{10}$	*Step 2.*	7.6 has one decimal place. One decimal place is tenths. Write 10 as the denominator.
$7\frac{6}{10} = 7\frac{3}{5}$	*Step 3.*	Reduce $7\frac{6}{10}$ by 2.

Change each decimal or mixed decimal to a fraction or mixed number. Reduce each fraction. Answers are on page 101.

1.	.04 =	.5 =	7.36 =	15.625 =
2.	.125 =	.008 =	1.0004 =	28.14 =
3.	.0035 =	.425 =	3.85 =	16.66 =
4.	.64 =	.75 =	9.16 =	34.002 =
5.	.065 =	.00075 =	10.024 =	8.00006 =

Changing Fractions to Decimals

A fraction is an instruction to divide. The line separating the numerator from the denominator means "divided by." For example, $\frac{1}{4}$ means 1 divided by 4. To change a fraction to a decimal, divide the numerator by the denominator. Put a decimal point and zeros to the right of the numerator.

EXAMPLE: Change $\frac{1}{4}$ to a decimal.

$$\begin{array}{r} .25 \\ 4\overline{)1.00} \\ \underline{8} \\ 20 \\ \underline{20} \end{array}$$

Step 1. Divide 1 by 4.

Step 2. Put a decimal point and two zeros to the right of 1.

Step 3. Divide and bring the decimal point up into the answers above its position in the problem.

Sometimes you can put just one zero to the right of the decimal point and the division will come out even. Sometimes you can keep putting zeros in the problem and the division will never come out even.

EXAMPLE: Change $\frac{5}{6}$ to a decimal.

Step 1. Divide 5 by 6.

Step 2. Put a decimal point and two zeros to the right of 5.

$$\begin{array}{r} .83\frac{2}{6} \\ 6\overline{)5.00} \\ \underline{4\,8} \\ 20 \\ \underline{18} \\ 2 \end{array}$$

Step 3. Divide and bring the decimal point up into the answer above its position in the problem. Notice that if we put more zeros in the problem, the division does not come out even. You can stop with two zeros.

Step 4. Make a fraction with the remainder 2 over the divisor 6.

$.83\frac{2}{6} = .83\frac{1}{3}$ *Step 5.* Reduce the fraction by 2.

Change each fraction to a decimal. Answers are on page 101.

1. $\frac{3}{4} =$ $\frac{2}{5} =$ $\frac{3}{10} =$ $\frac{1}{3} =$

2. $\frac{1}{6} =$ $\frac{7}{20} =$ $\frac{1}{2} =$ $\frac{5}{8} =$

3. $\frac{8}{25} =$ $\frac{5}{9} =$ $\frac{9}{10} =$ $\frac{1}{12} =$

4. $\frac{2}{7} =$ $\frac{1}{8} =$ $\frac{17}{25} =$ $\frac{2}{3} =$

5. $\frac{4}{5} =$ $\frac{5}{16} =$ $\frac{7}{8} =$ $\frac{1}{20} =$

6. $\frac{13}{20} =$ $\frac{13}{16} =$ $\frac{4}{7} =$ $\frac{7}{9} =$

7. $1\frac{1}{4} =$ $\frac{3}{8} =$ $\frac{9}{25} =$ $\frac{7}{50} =$

8. $\frac{7}{10} =$ $1\frac{2}{3} =$ $\frac{19}{20} =$ $\frac{1}{5} =$

9. $\frac{23}{50} =$ $\frac{24}{25} =$ $1\frac{11}{20} =$ $\frac{4}{9} =$

10. $\frac{1}{10} =$ $\frac{31}{50} =$ $\frac{3}{20} =$ $3\frac{3}{10} =$

Addition of Decimals

Adding decimals is one of the easiest operations in mathematics. To add decimals, line up the decimals with the decimal points under each other.

EXAMPLE: Add .56 + .3 + .418

Step 1. Line up the numbers with the decimal points under each other.

Step 2. Add each column. In the thousandths column the only digit is 8. In the hundredths column the digits are 6 and 1. In the tenths column the digits are 5, 3, and 4.

```
    .56
    .3
+   .418
  ------
   1.278
```

In this example the total of the tenths colums is 12. Only one digit can fit in each column. Write the 2 in the tenths column and carry the 1 over to the units column.

Add each problem. Answers are on page 102.

1. .37 + .2 + .608 =   .65 + .838 + .9 =

2. .4 + .9 + .7 =   .19 + .86 + .07 =

3. .49 + .8 + .784 =   .7 + .3266 + .285 =

4. .06 + .602 + .6 =   .355 + .5 + .0305 =

5. .852 + .0078 + .04 =   .4326 + .14 + .073 =

6. .009 + .28 + .7 =   .035 + .119 + .206 =

When you add whole numbers with decimals or mixed decimals, remember to put a point at the right of the whole number. Then line up the numbers with the decimal points under each other.

EXAMPLE: Add 3.62 + 18 + .475

```
   3.62
  18.
+   .475
--------
  22.095
```

Step 1. Put a decimal point at the right of 18.

Step 2. Line up the numbers with the decimal points under each other.

Step 3. Add each column.

7. 3.2 + 55 + 2.86 = .604 + 4.5 + 3 =

8. 90 + 5.134 + .37 = 8 + .73 + .626 =

9. .317 + 4 + .74 = 3.55 + .42 + 10 =

10. 3.496 + 2.85 + 7 = 12.084 + 19 + .36 =

11. 53 + 19.6 + 8.49 = .064 + 1.3 + 2 =

12. 1.0075 + 3 + .48 = 6 + .0793 + 1.8 =

13. 128.2 + 3.884 + 7 = .509 + 6.66 + 14 =

Adding Decimals Applications

These problems give you a chance to apply your skills in adding decimals. Watch for words like **sum** and **total**. They usually mean to add. Other words like **combine, complete, entire,** and **altogether** sometimes mean to add.

For each problem give your answer the correct label such as pounds or miles. Answers are on page 102.

1. Susan bought 4.5 pounds of ground beef, 3.8 pounds of chicken, and 3 pounds of pork from the butcher. What was the total weight of the meat?

2. In 1970 there were 40.7 million color televisions in the world. In 1980 there were 94.6 million more than in 1970. How many color T.V.s were there in 1980?

3. The reading on the mileage gauge in John's car was 29,342.6 miles on Monday morning. Monday he drove 248.7 miles. What was the reading on the gauge Monday night?

4. At the last census there were 3.4 million people in Chicago. There were 3.3 million people living in the suburbs around Chicago. What was the combined population of Chicago and its suburbs?

5. The average temperature in April in Detroit is 46.7°. The average temperature in April in Dallas is 18.5° higher than in Detroit. What is the average temperature in Dallas in April?

6. In 1970 there were 8.6 million people living in Mexico City. Experts believe in 1990 there will be 13 million more people in Mexico City than in 1970. How many people will be living in Mexico City in 1990?

7. Silvia packed three boxes of presents to send to her relatives. One box weighed 2.6 kilograms. Another weighed 3.135 kilograms. The third weighed 4.05 kilograms. What was the combined weight of the three boxes?

8. On Thursday David drove 2.4 miles to take his children to school. Then he drove 12.9 miles to get to work. After work he drove 9 miles to attend an evening class. After his class he drove 8.6 miles home. How many miles did David drive on Thursday?

9. In 1978 the U.S. government spent about $2.8 billion for public transportation systems, $2.3 billion for airports, and about $8 billion for highways. What amount did the government spend altogether?

10. Sam works part-time at a gas station. Friday he worked 5.5 hours. Saturday he worked 7.25 hours. Sunday he worked 4 hours. What total number of hours did Sam work that weekend?

11. In 1970 the price of an ounce of gold was $35.20. In 1979 an ounce was worth $473.55 more than in 1970. What was the price of an ounce of gold in 1979?

Subtraction of Decimals

To subtract decimals, line up the decimals with the points under each other, just like in addition. Remember to put a point at the right of a whole number. Put zeros at the right until each decimal has the same number of places. You will need the zeros for borrowing.

EXAMPLE: Subtract 6 − .24

```
   6.
 −  .24
 -------
```

Step 1. Put a decimal point at the right of 6.

Step 2. Line up the numbers with the points under each other.

```
  5 9
  6.00
 −  .24
 -------
  5.76
```

Step 3. Put two zeros at the right of 6 to give each decimal the same number of places.

Step 4. Borrow and subtract.

Subtract each problem. Answers are on page 102.

1.	5 − .248 =	10 − .15 =	.8 − .563 =
2.	.49 − .087 =	3 − .068 =	.01 − .003 =
3.	2.5 − 1.836 =	4.7 − .266 =	.64 − .305 =
4.	12 − 6.98 =	3 − 1.064 =	9.76 − 2 =
5.	.08 − .063 =	20 − .9 =	.2 − .038 =
6.	.025 − .009 =	1 − .3407 =	12 − .3 =
7.	9.8 − 2.72 =	.4 − .369 =	5 − .505 =

Subtracting Decimals Applications

These problems give you a chance to apply your skills in subtracting decimals. Watch for words like **difference** and **balance.** They usually mean to subtract. Phrases like **how much more** and **how much less** also mean to subtract. Remember to put the bigger number on top.

Give your answers the correct labels such as meters or pounds. Answers are on page 102.

1. In 1970 there were 104.2 million females in the U.S. and 98.9 million males. How many more females than males were there?

2. The Midvale Community Organization hopes to raise $2 million to build a new recreation center. So far they have $1.25 million. How much more money do they need?

3. In 1790 there were 4.5 people for every square mile of land in the U.S. In 1975 there were 60.3 people for every square mile. The number of people for every square mile was how much less in 1790 than in 1975?

4. Petra is 1.7 meters tall. Her daughter Rachel is 1.25 meters tall. How much taller is Petra than her daughter?

5. In 1969 the average American ate 46.7 pounds of chicken during the year. In 1979 the average was 62 pounds per person. How many more pounds of chicken did the average American eat in 1979 than in 1969?

6. Babe Ruth's batting average for his career was .342. Ted Williams' batting average was .344. How much better was Ted Williams' average than Babe Ruth's?

7. The Atlas Steel Company produced 66.9 tons of steel in 1970, and 79.7 tons of steel in 1980. How much more steel did Atlas produce in 1980?

8. Phil made $63.50 washing cars. He spent $12.90 on a new shirt. How much money did Phil have left?

9. In 1978 Canada produced 1.6 million barrels of oil a day. Canada used 1.9 million barrels of oil a day in 1978. How much less oil did Canada produce than it used in 1978?

10. The express bus takes 15.8 minutes to get from Jackson Street to First Street. The local bus takes 21.6 minutes for the same trip. Find the difference in minutes between the express bus and the local bus.

11. In 1800 the population of the United States was 5.3 million. In 1900 the population reached 75.9 million. How much did the population of the United States increase between 1800 and 1900?

12. Ken had $112.45 in a savings account. He took $48.30 out of his account to pay for a new tire. What is the new balance in Ken's savings account?

More Subtracting Decimals Applications

Use the bar graphs on this page to answer the next questions. In some of the problems you will need to add. Read each question carefully. Then decide which operation to use. Answers are on page 102.

Amount of Oil Produced in Billions of Barrels

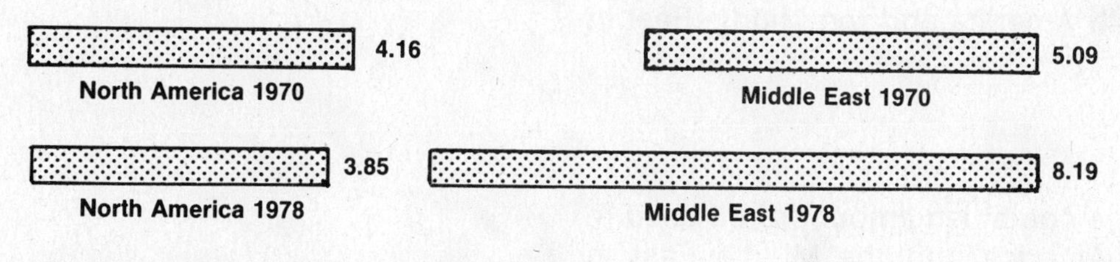

Amount of Oil Used in Billions of Barrels

6.09
North America 1970
.39
Middle East 1970
7.68
North America 1978
.68
Middle East 1978

1. Find the combined amount of oil produced in North America and the Middle East in 1970.

2. Find the combined amount of oil used in North America and the Middle East in 1970.

3. The amount of oil produced in both North America and the Middle East in 1970 was how much more than the amount used in these two areas that year?

4. How much less oil did North America produce in 1978 than in 1970?

5. How much more oil did the Middle East produce in 1978 than in 1970?

6. How much more oil did North America use in 1978 than it produced in 1978?

7. How much less oil did the Middle East use in 1978 than it produced in 1978?

8. Find the combined amount of oil produced in North America and the Middle East in 1978.

9. Find the combined amount of oil used in North America and the Middle East in 1978.

10. The amount of oil produced in both North America and the Middle East in 1978 was how much more than the amount used in these two areas in the same year?

11. How much more oil did North America use in 1978 than in 1970?

12. How much more oil did the Middle East use in 1978 than in 1970?

13. How much more oil did North America use in 1970 than it produced in 1970?

14. How much less oil did the Middle East use in 1970 than it produced in 1970?

Multiplication of Decimals

Multiplying decimals is easier than adding or subtracting decimals. You do not have to line up the decimals. Just count the number of decimal places in each number you multiply. Put the total number of places from the two numbers in the answer.

EXAMPLE: Multiply 1.47 × .3

Step 1. Set the numbers up for easy multiplication. Since .3 has only one digit, put it below.

Step 2. Multiply the numbers.

```
   1.47
×    .3
  .441
```

Step 3. Count the number of decimal places in each number. 1.47 has two decimal places. .3 has one decimal place.

Step 4. Put the total number of decimal places (2 + 1 = 3) in the answer.

Remember that a whole number has no decimal places.

EXAMPLE: Multiply 7 × 4.6

Step 1. Set the numbers up for easy multiplication.

Step 2. Multiply the numbers.

```
   4.6
×    7
  32.2
```

Step 3. Count the number of decimal places in each number. 4.6 has one decimal place. 7 has no decimal places.

Step 4. Put the total number of decimal places (0 + 1 = 1) in the answer.

Sometimes you will need to put extra zeros in your answer.

EXAMPLE: Multiply .2 × .09

Step 1. Set the numbers up for easy multiplication.

Step 2. Multiply the numbers.

```
   .09
×   .2
  .018
```

Step 3. Count the number of decimal places in each number. .2 has one decimal place. .09 has two decimal places.

Step 4. Put the total number of decimal places (1 + 2 = 3) in the answer. Put a zero at the left of 18 to make three places.

Sometimes you can take zeros off your answer.

EXAMPLE: Multiply 4.8 × .05

Step 1. Set the numbers up for easy multiplication.

Step 2. Multiply the numbers.

$$\begin{array}{r} 4.8 \\ \times \ \ .05 \\ \hline .24\not{0} \end{array}$$

Step 3. Count the number of decimal places in each number. 4.8 has one decimal place. .05 has two decimal places.

.24

Step 4. Put the total number of decimal places (1+2=3) in the answer.

Step 5. Take off the last zero from the answer. The zero is not necessary. It does not change the value of the other digits. 2 is still in the tenths place. 4 is still in the hundredths place.

Multiply each problem. Answers are on page 103.

1. 6 × 1.8 = .73 × 8 = 9 × .04 =

2. .08 × .7 = .4 × 9.2 = .36 × .5 =

3. .6 × 7.12 = .863 × .8 = .3 × 75.4 =

4. .12 × 73 = 2.8 × 6.5 = .87 × 6.6 =

5. .038 × .4 = .009 × .05 = .24 × .008 =

6. $56 \times 7.7 =$ $.63 \times 6.5 =$ $2.7 \times 3.6 =$

7. $.26 \times .025 =$ $.041 \times 3.9 =$ $6.4 \times 8.01 =$

8. $426 \times .05 =$ $2.3 \times 187 =$ $349 \times .18 =$

9. $5.62 \times 6.4 =$ $.749 \times 2.3 =$ $23.4 \times .48 =$

10. $1.5 \times 3.17 =$ $.91 \times 80.6 =$ $4.6 \times .625 =$

11. $.037 \times .02 =$ $.16 \times .009 =$ $.036 \times 2.4 =$

12. $2.38 \times 40 =$ $70 \times .309 =$ $67.5 \times 80 =$

13. $.037 \times .56 =$ $928 \times .07 =$ $.0925 \times .4 =$

Multiplication of Decimals by 10, 100, 1000

There are short ways to multiply decimals by 10, 100, or 1000.

When you multiply a decimal by 10, move the decimal point one place to the right.

EXAMPLE: $.36 \times 10 = .3_{\smile}6 = 3.6$

When you multiply a decimal by 100, move the decimal point two places to the right.

EXAMPLE: $100 \times 2.8 = 2.80_{\smile} = 280$

Notice that a zero is placed to the right of 2.8 to move two places. Also notice that the final decimal point is not necessary in this whole number answer.

When you multiply a decimal by 1000, move the decimal point three places to the right.

EXAMPLE: $.2305 \times 1000 = .230_{\smile}5 = 230.5$

Write the answers to each problem. Answers are on page 104.

1. $.4 \times 10 =$ $\qquad$ $1.7 \times 10 =$ $\qquad$ $.28 \times 10 =$ $\qquad$ $1.48 \times 10 =$

2. $10 \times .925 =$ $\qquad$ $10 \times 66.7 =$ $\qquad$ $10 \times .08 =$ $\qquad$ $10 \times .037 =$

3. $.62 \times 100 =$ $\qquad$ $4.53 \times 100 =$ $\qquad$ $.129 \times 100 =$ $\qquad$ $5.7 \times 100 =$

4. $100 \times 4.02 =$ $\qquad$ $100 \times .58 =$ $\qquad$ $100 \times .344 =$ $\qquad$ $100 \times 16.6 =$

5. $1.485 \times 1000 =$ $\qquad$ $.36 \times 1000 =$ $\qquad$ $23.5 \times 1000 =$ $\qquad$ $.705 \times 1000 =$

6. $1000 \times 2.539 =$ $\qquad$ $1000 \times .085 =$ $\qquad$ $1000 \times 3.6 =$ $\qquad$ $1000 \times 42.12 =$

Multiplying Decimals Applications

These problems give you a chance to apply your skills in multiplying decimals. Some problems give you the weight of one thing or the cost of one thing. Then you have to find the weight or cost of several things. These situations mean to multiply.

Give every answer the correct label such as grams or dollars. Answers are on page 104.

1. Manny makes $5.60 an hour. Last week he worked 38.5 hours. How much money did Manny make last week?

2. Gordon's car gets an average of 26.7 miles of city driving on a gallon of gasoline. How many miles can Gordon drive in the city on 8 gallons of gasoline?

3. Celeste weighs 130 pounds. One pound equals 0.45 kilograms. What is Celeste's weight in kilograms?

4. The distance from Pete's house to the factory where he works is 6.5 miles. One mile equals 1.6 kilometers. What is the distance from Pete's house to his factory in kilometers?

5. In a recent year an average person ate 13.7 pounds of fish. An average family of 5 people ate how many pounds of fish that year?

6. Cheddar cheese costs $2.30 a pound. Find the cost of 2.5 pounds of cheddar cheese.

7. Silvia is 66 inches tall. One inch equals 2.54 centimeters. What is Silvia's height in centimeters?

8. When Tom works overtime he makes $10.50 an hour. Last week Tom worked 8.5 hours overtime. How much did Tom make for overtime work last week?

9. Pork chops cost $1.50 a pound. Find the cost of 3.3 pounds of pork chops.

10. A train travels at an average speed of 52 miles per hour. How far can it go in 4.75 hours?

11. One can of soup weighs .27 kilogram. What is the weight of 12 cans of soup?

12. It costs $.025 to run a black and white television for one hour. How much does it cost to run a television for 40 hours?

13. Matt bought 6.75 feet of lumber. The lumber cost $3.80 a foot. What was the total cost of the lumber?

14. Frank wants to build a square picture frame. He needs 15.2 inches of framing for each side. Find the total length of framing Frank needs to put around all four sides of the picture.

More Multiplying Decimals Applications

Use the chart below to answer the following questions. The chart is about railroad travel in the U.S. Answers are on page 104.

Year	Average speed of trains in miles per hour	Average time of a ride in hours	Total number of riders in billions
1960	40.7	1.6	.33
1965	41.3	1.41	.30
1970	41.0	1.05	.28
1975	39.2	.92	.27

1. Find the average distance each rider traveled. Multiply the average speed by the average time of a ride. Answers will be in miles.

1960 ____________

1965 ____________

1970 ____________

1975 ____________

2. The average distance for a ride on the railroad in 1975 was how much less than the average distance in 1960?

3. Change the average distance each rider traveled from miles to kilometers. To change miles to kilometers, multiply the number of miles by 1.6.

1960 ____________

1965 ____________

1970 ____________

1975 ____________

4. Find the total distance all riders traveled in a year. To find the total distance, multiply the total number of riders by the average distance each rider traveled. Answers will be in billions of miles.

1960 ____________

1965 ____________

1970 ____________

1975 ____________

5. The total distance traveled by all riders in 1965 was how much more than the distance traveled by all riders in 1970?

6. Change the number of riders to millions. Multiply the total number of riders in billions by 1000. Answers will be in millions of riders.

1960 ____________

1965 ____________

1970 ____________

1975 ____________

7. How many more riders used trains in 1960 than in 1975?

8. How many more riders used trains in 1970 than in 1975?

9. In 1975 passengers paid about $.06 per mile for travel on railroads. If the average ride in 1975 was about 36 miles, how much did a passenger pay for the average ride in 1975?

Division of Decimals by Whole Numbers

You began to divide decimals when you changed fractions into decimals on page 14. There you divided a whole number by a bigger whole number. When you divide a decimal by a whole number, line up your problem carefully. Then bring the decimal point up into the answer above its position in the problem.

EXAMPLE: Divide 5.36 ÷ 8

$$\begin{array}{r} .67 \\ 8\overline{)5.36} \\ \underline{4\,8} \\ 56 \\ \underline{56} \end{array}$$

Step 1. Set the problem up for long division.

Step 2. Divide.

Step 3. Bring the decimal point up into the answer above its position in the problem.

Sometimes you will need to put zeros in your answer.

EXAMPLE: Divide .364 ÷ 7

$$\begin{array}{r} .052 \\ 7\overline{).364} \\ \underline{35} \\ 14 \\ \underline{14} \end{array}$$

Step 1. Set the problem up for long division.

Step 2. Divide.

Step 3. Bring the decimal point up into the answer above its position in the problem. To show that 7 does not divide into .3, put a zero above the 3.

Divide each problem. Answers are on page 105.

1. 33.6 ÷ 7 = 5.67 ÷ 9 = .116 ÷ 4 =

2. 14.28 ÷ 6 = 4.215 ÷ 5 = 1.71 ÷ 3 =

3. .182 ÷ 13 = 140.4 ÷ 36 = .392 ÷ 14 =

4. 543.6 ÷ 18 = 1.804 ÷ 22 = 30.24 ÷ 48 =

5. $498.4 \div 4 =$ $210.63 \div 7 =$ $483.69 \div 23 =$

6. $8.127 \div 9 =$ $.493 \div 17 =$ $17.28 \div 8 =$

7. $.0156 \div 39 =$ $703.2 \div 16 =$ $4.56 \div 12 =$

8. $20.664 \div 24 =$ $78.003 \div 3 =$ $6.136 \div 13 =$

9. $51.8 \div 37 =$ $33.58 \div 46 =$ $1.4286 \div 3 =$

10. $8.91 \div 11 =$ $269.8 \div 38 =$ $1776.61 \div 31 =$

11. $186.9 \div 89 =$ $10.534 \div 458 =$ $22.1 \div 17 =$

12. $7.79 \div 19 =$ $410.3 \div 11 =$ $3211.6 \div 37 =$

Division of Decimals by Decimals

Dividing by decimals is the most complicated decimal operation. First change the problem to a new problem. In the new problem the number you are dividing by (the divisor) should be a whole number. You can change the divisor to a whole number by moving the decimal point to the right end. Then move the decimal point in the other number (the dividend) the same number of places.

These steps are easier to understand with whole numbers. Look at the problem $10 \div 2 = 5$. The answer is the same if we move the decimal point one place to the right in both 10 and 2: $100 \div 20 = 5$. The problems are different but the answers are the same.

With decimal division you can also change the problem and get the right answer. The object is always to make the divisor a whole number.

Follow these examples carefully.

EXAMPLE: Divide $3.48 \div .6$

$.6\overline{)3.48}$ *Step 1.* Set the problem up for long division.

$.6\overline{)3.48}$ *Step 2.* Move the decimal point in the divisor .6 one place to the right to make it a whole number.

Step 3. Also move the decimal point in the dividend 3.48 one place to the right. Note that the decimal point in the divisor and the dividend are always moved the same number of places to the right.

$$\begin{array}{r} 5.8 \\ .6\overline{)3.48} \\ 30 \\ \hline 48 \\ 48 \\ \hline \end{array}$$

Step 4. Divide and bring the decimal point up into the answer above its new position.

It is a good idea to check division of decimals problems. Multiply the answer by the divisor. The result should equal the dividend. For the example above:

$$\begin{array}{r} 5.8 \\ \times \quad .6 \\ \hline 3.48 \end{array}$$

Sometimes you will have to put extra zeros with the dividend.

EXAMPLE: Divide 2.7 ÷ .03

$.03\overline{)2.7}$ *Step 1.* Set the problem up for long division.

$.03\overline{)2.7}$ *Step 2.* Move the decimal point in the divisor .03 two places to the right to make it a whole number.

Step 3. Also move the decimal point in the dividend 2.7 two places to the right. Put an extra zero to the right of 2.7 to get two decimal places.

```
     90.
.03)2.70
    2 7
    ---
      0
```

Step 4. Divide and bring the decimal point up into the answer above its new position.

To check this example, multiply 90 by .03:

```
    90
 ×  .03
 ------
  2.70 = 2.7
```

(In the product, the final 0 of 2.70 is crossed out.)

Divide and check each problem. Answers are on page 105.

1. 2.76 ÷ .6 =   .256 ÷ .8 =   6.03 ÷ .9 =

2. .236 ÷ .4 =   2.82 ÷ .3 =   23.1 ÷ .7 =

3. 5.67 ÷ .09 =   .294 ÷ .06 =   .0768 ÷ .08 =

4. .0174 ÷ .03 =   .595 ÷ .07 =   18.8 ÷ .04 =

5. .0215 ÷ .05 =   .312 ÷ .08 =   6.03 ÷ .09 =

6. .28 ÷ .008 = .329 ÷ .007 = .232 ÷ .004 =

7. .0159 ÷ .003 = 15.66 ÷ .006 = .1206 ÷ .009 =

8. 9.3 ÷ .15 = 31.54 ÷ .38 = 1.767 ÷ 1.9 =

9. 1.118 ÷ 2.6 = 3.328 ÷ .52 = 30.75 ÷ 4.1 =

10. 4.68 ÷ .018 = 233.06 ÷ .043 = 27.72 ÷ .072 =

11. 17.36 ÷ .035 = 83.6 ÷ .011 = 11.27 ÷ .023 =

12. 1.29 ÷ 21.5 = 2.41 ÷ .482 = 245.6 ÷ 3.07 =

13. .12077 ÷ 9.29 = 36.624 ÷ 87.2 = 42.28 ÷ .604 =

Division of Whole Numbers by Decimals

To divide a whole number by a decimal, remember to put a decimal point at the right of the whole number. Then move the points in both the divisor and the dividend. You will have to put zeros with the dividend.

Follow this example carefully.

EXAMPLE: Divide 12 ÷ .004

.004)12.

Step 1. Set the problem up for long division. Put a decimal point at the right of 12.

.004)12.

Step 2. Move the decimal point in the divisor .004 three places to the right to make it a whole number.

```
     3 000.
.004)12.000
     12
      0 000
```

Step 3. Also move the decimal point in the dividend 12 three places to the right. Put three zeros to the right of 12 to get three decimal places.

Step 4. Divide and bring the decimal point up into the answer above its new position.

To check this example, multiply 3000 by .004:

```
   3000
×  .004
12.000 = 12
```

Divide and check each problem. Answers are on page 105.

1.	18 ÷ 1.2 =	20 ÷ .25 =	33 ÷ .04 =
2.	42 ÷ 3.5 =	8 ÷ .02 =	72 ÷ .024 =
3.	35 ÷ .14 =	28 ÷ .16 =	10 ÷ .025 =
4.	108 ÷ 1.35 =	18 ÷ .075 =	264 ÷ 1.6 =

5. $2485 \div .35 =$ $351 \div 1.3 =$ $1400 \div .05 =$

6. $251 \div .004 =$ $3416 \div 42.7 =$ $814 \div 2.2 =$

7. $18 \div 1.5 =$ $246 \div .6 =$ $1079 \div 8.3 =$

8. $1140 \div .76 =$ $8 \div .04 =$ $4847 \div 13.1 =$

9. $118{,}406 \div 81.1 =$ $588 \div 4.2 =$ $5751 \div 21.3 =$

10. $2635 \div 3.1 =$ $3640 \div .065 =$ $612 \div 1.8 =$

Division of Decimals by 10, 100, 1000

There are short ways to divide decimals by 10, 100, or 1000.

When you divide a decimal by 10, move the decimal point one place to the left.

EXAMPLE: .5 ÷ 10 = 0.5 = .05

Notice that a zero is placed to the left of .5.

When you divide a decimal by 100, move the decimal point two places to the left.

EXAMPLE: 456.3 ÷ 100 = 4.56.3 = 4.563

When you divide a decimal by 1000, move the decimal point three places to the left.

EXAMPLE: 8.7 ÷ 1000 = .008.7 = .0087

Write the answers to each problem. Answers are on page 105.

1.	4.3 ÷ 10 =	18 ÷ 10 =	.6 ÷ 10 =	.034 ÷ 10 =
2.	3.28 ÷ 10 =	9 ÷ 10 =	17.8 ÷ 10 =	.15 ÷ 10 =
3.	19.2 ÷ 100 =	1.87 ÷ 100 =	21 ÷ 100 =	8 ÷ 100 =
4.	.47 ÷ 100 =	6.5 ÷ 100 =	.029 ÷ 100 =	364 ÷ 100 =
5.	28.4 ÷ 1000 =	4.39 ÷ 1000 =	350 ÷ 1000 =	2.6 ÷ 1000 =
6.	3,980 ÷ 1000 =	.9 ÷ 1000 =	42.16 ÷ 1000 =	923.5 ÷ 1000 =

Dividing Decimals Applications

These problems give you a chance to apply your skills in dividing decimals. Some problems give you the weight of several things or the cost of several things. Then you have to find the weight or cost of one thing. These problems mean to divide. Divide the weight or cost of the things by the number of things.

EXAMPLE: Luba bought 4.5 pounds of chicken for $3.24. Find the price of one pound of chicken.

4.5)$3.24

Step 1. Divide the price of several pounds ($3.24) by the number of pounds (4.5).

4.5)$3.24

Step 2. Move the decimal point in the divisor 4.5 one place to the right to make it a whole number.

```
     $  .72
4.5)$3.2 40
     3 1 5
        90
        90
```

Step 3. Also move the decimal point in the dividend $3.24 one place to the right.

Step 4. Divide and bring the decimal point up into the answer above its new position. Put an extra zero at the right of 32.4 to give the answer the two decimal places for money.

The words **cutting, sharing,** and **splitting** usually mean to divide.

EXAMPLE: Mark wants to split a 1.95 meter board into 3 equal pieces. How long will each piece be?

```
   .65 meters
3)1.95 meters
  1 8
    15
    15
```

Step 1. Divide the length of the board by 3.

Step 2. Bring the decimal point up into the answer above its position in the dividend.

A problem that asks how many times a decimal fits into another number also means to divide.

EXAMPLE: There are 2.54 centimeters in an inch. How many inches are there in 12.7 centimeters?

2.54)12.7	*Step 1.*	Find out how many times 2.54 goes into 12.7. Divide 12.7 by 2.54.
2.54)12.7	*Step 2.*	Move the decimal point in the divisor 2.54 two places to the right.
5. 2.54)12.70 12 70	*Step 3.*	Also move the decimal point in the dividend 12.7 two places to the right. Put a zero to the right of 12.7 to get two decimal places.
	Step 4.	Divide and bring the decimal point up into the answer above its new position.
5 inches		

Give each answer the correct label such as dollars or meters. Answers are on page 106.

1. Noreen bought 3.4 pounds of ground beef for $5.44. What was the price of one pound of the ground beef?

2. Jeff shipped a crate weighing 85 pounds. He paid $52.70 to ship the crate. What was the shipping price for each pound?

3. A box contains 12 cans of tomato paste. The total weight of all the cans in the box is 9.48 kilograms. How much does one can of tomato paste weigh?

4. Pete is a plumber. He wants to cut 2.52 meters of copper tube into 6 equal pieces. How long will each piece be?

5. The Jackson family picked 53.6 pounds of apples. They want to share the apples equally with three other families and themselves. How many pounds of apples will each of the four families get?

6. The cost of a new gymnasium for Center City is $1.95 million. The cost will be split equally with a grant from the state, a loan, and gifts. Find the amount of each of the three parts.

7. There are 2.2 pounds in one kilogram. Sally weighs 132 pounds. What is her weight in kilograms?

8. There are .62 miles in a kilometer. How many kilometers are there in 46.5 miles?

9. There are 3.8 liters in a gallon. How many gallons are there in 152 liters?

10. John drove 222.3 miles on 9.5 gallons of gasoline. How far did John drive on one gallon of gas?

11. Susan makes $4.60 an hour. Last week she made $177.10 before taxes. How many hours did Susan work last week?

12. The height of Mount Everest is 8,848 meters. What is the height of Mount Everest in kilometers?*(To change meters to kilometers, divide by 1000).*

More Dividing Decimals Applications

Use the chart below to answer the following questions. The chart shows the number of miles four different men drove in one week. It also shows the numbers of gallons of gas each man bought. The four men want to compare the miles each of their cars can travel on one gallon of gasoline. Answers are on page 106.

Name	Miles driven in one week	Gallons of gas bought in one week	Number of miles per gallon
John	289.8	25.2	________
Mark	1020.9	41.5	________
Pete	190.8	12	________
Dave	1155.4	54.5	________

1. Find the number of miles each man's car gets on one gallon of gas. To find gas mileage, divide the number of miles each man drove in one week by the number of gallons each man bought in one week. Answers will be in miles per gallon.

   John ____________

   Mark ____________

   Pete ____________

   Dave ____________

2. Who drove the most that week?

3. Who drove the least that week?

4. Who got the most miles per gallon?

5. Who got the fewest miles per gallon?

6. What was the difference between the most and the fewest miles per gallon?

The list below tells how much each man spent for gas the week they compared their mileage.

John	$26.46
Mark	$45.65
Pete	$12.96
Dave	$61.04

7. Find out how much each man paid for a gallon of gas. Divide the amount each paid in a week for gas by the number of gallons each man bought.

   John ____________

   Mark ____________

   Pete ____________

   Dave ____________

8. Who paid the most for a gallon of gas?

9. Who paid the least for a gallon of gas?

10. Find the number of miles the four men drove altogether that week.

11. Find the total number of gallons of gas the four men bought that week.

12. Find the total amount of money the four men spent on gasoline that week.

13. Find the difference between the most and the least any of the four men spent on gas for the week.

Decimal Review

These problems will tell you which parts of the decimal section of this book you need to review before going on to percents. When you finish, check the chart to see which pages you need to review.

1. Circle the mixed decimals in this list.

5.18 465 .3306 4.7 .083

2. For each decimal or mixed decimal write how many decimal places the number has.

a. 79.4 ______ .885 ______ 40.32 ______

b. .0371 ______ 8.68 ______ 3.927 ______

3. Cross out the zeros that are not needed in each decimal or mixed decimal.

60.040 08.0050 300.2900 007.0430

4. Write these decimals or mixed decimals in words.

a. .8 ______ 9.35 ______

b. .056 ______ 12.01 ______

c. .0103 ______ 30.226 ______

d. .00325 ______ 2.000009 ______

5. Write each number as a decimal or a mixed decimal.

a. eight and fourteen hundredths ______

b. thirty-nine ten-thousandths ______

c. one hundred five and seven thousandths ______

d. four thousand seven hundred eight hundred-thousandths______

e. seventy-two and six hundredths ______

6. Circle the bigger decimal in each pair.

a. .4 or .39   .018 or .01   .7 or .27

b. .63 or .605   .05 or .50   .14 or .041

7. Circle the biggest decimal in each group.

a. .67, .607, or .76   .202, .22, or .2   .4, .346, or .46

b. .005, .051, or .05   .17, .071, or .107   .32, .305, or .3

8. Change each decimal or mixed decimal to a fraction or mixed number. Reduce each fraction.

a. .7 =   4.25 =   6.24 =   .00005 =

b. .0048 =   .875 =   .018 =   12.32 =

9. Change each fraction to a decimal.

a. $\frac{7}{20} =$ $\frac{3}{7} =$ $\frac{11}{12} =$ $\frac{4}{25} =$

b. $\frac{5}{16} =$ $\frac{1}{9} =$ $\frac{9}{50} =$ $\frac{1}{8} =$

10. .3 + .8 + .6 =

11. .67 + .8 + .616 =

12. .0029 + .32 + .085 =

13. .33 + .417 + .28 =

14. 3.35 + 17 + . 086 =

15. .42 + 9.68 + 12 =

16. $7.24 + .035 + 8.6 =$

17. $29 + 2.88 + .046 =$

18. In 1970 a Swiss franc was worth 22.87¢ in U.S. money. In 1979 a Swiss franc was worth 44.83¢ more than in 1970. What was the Swiss franc worth in 1979?

19. The porch in front of George and Martha's house was 4.5 meters long. George built an extension that makes the porch 2.75 meters longer. How long is the porch now?

20. $.8 - .296 =$

21. $9.1 - .36 =$

22. $2.5 - 1.482 =$

23. $8 - .16 =$

24. $4 - .037 =$

25. $7.6 - 2.47 =$

26. The distance from Joe's house to his factory is exactly 9 miles. Joe stopped to get gas when he was 3.4 miles away from home. How many miles is it from the gas station to the factory?

27. By June the town of Midvale raised $.85 million for a new sports center. They need $1.3 million to build the center. How much more money do they need?

28. $9 \times .326 =$

29. $9.3 \times 4.5 =$

30. $6.8 \times .42 =$

31. $9.3 \times .006 =$

32. $.0037 \times 13 =$

33. $2.4 \times .635 =$

34. $2.8 \times 10 =$

35. $100 \times .8 =$

36. $.4681 \times 1000 =$

37. Jack weighs 160 pounds. One pound equals .45 kilogram. What is Jack's weight in kilograms?

38. Dave makes $6.20 an hour. How much does he make for a day when he works 7.5 hours?

39. $39.2 \div 8 =$

40. $1.798 \div 29 =$

41. $3.42 \div .9 =$

42. $1.28 \div .08 =$

43. $13.52 \div .026 =$

44. $1.512 \div 3.6 =$

45. $.216 \div 2.4 =$

46. $121.9 \div .53 =$

47. $112 \div .28 =$

48. $444 \div 3.7 =$

49. $6.2 \div 10 =$

50. $17 \div 100 =$

51. $3.2 \div 1000 =$

52. Floria bought 4.3 pounds of chicken for $3.01. What was the price of one pound of chicken?

53. Al wants to split a board 2.92 meters long into 4 equal pieces. How long will each piece be?

Check your answers on page 106. Circle the problems you missed on the chart below. Review the pages that show how to work the problems you missed. Then try the problems again.

Problem Number	Review Pages	Problem Number	Review Pages
1	8 - 9	28	25 - 27
2	8 - 9	29	25 - 27
3	8 - 9	30	25 - 27
4	10	31	25 - 27
5	11	32	25 - 27
6	12	33	28
7	12	34	28
8	13	35	28
9	14	36	28
10	16	37	29 - 30
11	16	38	29 - 30
12	16	39	33
13	16	40	33
14	17	41	35 - 37
15	17	42	35 - 37
16	17	43	35 - 37
17	17	44	35 - 37
18	18 - 19	45	35 - 37
19	18 - 19	46	35 - 37
20	20	47	38
21	20	48	38
22	20	49	39
23	20	50	39
24	20	51	39
25	20	52	41 - 43
26	21	53	41 - 43
27	21		

Step One to Percent Skill

The problems on pages 52 through 55 will help you find out which parts of the percent section of this book you need to work on. Do all the problems you can. At the end of the problems look at the chart on page 56 to see which page you should go to next.

1. Change each decimal to a percent.

$.4 =$ $.04 =$ $.225 =$ $.06\frac{1}{4} =$

2. Change each percent to a decimal.

$30\% =$ $9\% =$ $3\frac{1}{3}\% =$ $250\% =$

3. Change each fraction to a percent.

$\frac{13}{20} =$ $\frac{7}{8} =$ $\frac{7}{16} =$ $\frac{9}{50} =$

4. Change each percent to a fraction.

$16\% =$ $4.5\% =$

$44\frac{4}{9}\% =$ $29\frac{1}{6}\% =$

5. Find 30% of 70.

6. Find 25% of 124.

7. Find $37\frac{1}{2}\%$ of 64.

8. Find $83\frac{1}{3}\%$ of 300.

9. Find 150% of 18.

10. Find 220% of 75.

11. Find 2.3% of 200.

12. Find 1.75% of 480.

13. The Johnsons make $14,000 a year. They spend 30% of their income for food. How much do they spend on food in a year?

14. In 1970 there were 230,000 people living in Central County. In 1975 the number of people in the county was 110% of the 1970 number. How many people lived in Central County in 1975?

15. Find 6% of $7.49. Round off to the nearest cent.

16. Find 3.8% of $4.56. Round off to the nearest cent.

17. Harry bought a pair of shoes for $19.50. The sales tax in his state is 8%. How much did Harry pay for the shoes, including the sales tax?

18. Valerie makes $210 a week. Her employer takes out 17% of her check for taxes and social security. How much does Valerie take home in a week?

19. Find the interest on $800 at 6% annual interest for one year.

20. Find the interest on $800 at 6.5% annual interest for one year and three months.

21. 14 is what % of 35?

22. 80 is what % of 120?

23. 36 is what % of 40?

24. 16 is what % of 50?

25. Last year Sam weighed 240 pounds. He went on a diet and lost 24 pounds. What percent of his weight did Sam lose?

26. Mary works 40 hours a week. She spends about 24 hours every week filing. What percent of her work week does Mary spend filing?

27. In 1970 there were 4 members in Mr. and Mrs. Miller's family. In 1980 there were 7 members. By what percent did the size of the family increase?

28. Frank bought new tires on sale for $85. Before the sale the tires cost $100. Find the percent of discount on the original price.

29. $33\frac{1}{3}$% of what number is 45?

30. 37.5% of what number is 36?

31. 75% of what number is 90?

32. 35% of what number is 49?

33. 75% of the members of a firemen's union voted to strike. 126 members voted to strike. How many members are there in the union?

34. Tom spends $105 a month on payments for his car loan. The payments are 15% of his income. How much does Tom make in a month?

Check your answers on page 107. Then complete the chart below.

Problem numbers	Number of problems in this section	Number of problems you got right in this section	
1 to 4	4	______	If you had fewer than 3 problems right, go to page 57.
5 to 12	8	______	If you had fewer than 6 problems right, go to page 67.
13 to 20	8	______	If you had fewer than 6 problems right, go to page 71.
21 to 28	8	______	If you had fewer than 6 problems right, go to page 81.
29 to 34	6	______	If you had fewer than 4 problems right, go to page 88.

What Is a Percent?

You already learned that decimals are a kind of fraction. Percents are also a kind of fraction. Percents are used in everyday business situations. Interest rates, sales tax, and discounts are all measured in percents. For example, 6% is the sales tax rate in some states. 6% means 6 out of 100 equal parts. For every 100 cents you spend on something, you pay 6 cents in tax.

Percents are different from common fractions in two ways. One difference is that 100 is the only number that can be a denominator for percents. The other difference is that the denominator 100 is not written. Instead of writing 100, we write the percent sign, %.

Percents are almost like two-place decimals. A decimal with two places is called hundredths. The denominator for percents is 100. Instead of two decimal places, percents use the percent sign.

The list below shows examples of percents and the decimals and fractions each percent is equal to. Study the list carefully to learn the differences.

Percent	Decimal	Fraction
25%	.25	$\frac{25}{100}=\frac{1}{4}$
6%	.06	$\frac{6}{100}=\frac{3}{50}$
3.4%	.034	$\frac{34}{1000}=\frac{17}{500}$
80%	.8	$\frac{8}{10}=\frac{4}{5}$
100%	1.	1
150%	1.5	$1\frac{5}{10}=1\frac{1}{2}$

Changing Decimals to Percents

It is easy to change decimals to percents. Move the decimal point two places to the **right** and write a percent sign.

EXAMPLE: Change .25 to a percent.

.25 = .25 = 25% *Step 1.* Move the decimal point two places to the right.

Step 2. Write a percent sign.

Notice that when the decimal point moves to the end, you do not have to write the point.

EXAMPLE: Change .043 to a percent.

.043 = .04 3 = 4.3% *Step 1.* Move the decimal point two places to the right.

Step 2. Write a percent sign.

EXAMPLE: Change $.16\frac{2}{3}$ to a percent.

$.16\frac{2}{3} = .16\frac{2}{3} = 16\frac{2}{3}\%$ *Step 1.* Move the decimal point two places to the right.

Step 2. Write a percent sign.

Notice that when the decimal point comes just before a fraction, you do not have to write the point.

Sometimes you will have to put zeros after the decimal to get two places.

EXAMPLE: Change .3 to a percent.

.3 = .30 = 30% *Step 1.* Put a zero at the right of .3

Step 2. Move the decimal point two places to the right.

Step 3. Write a percent sign.

Change each decimal to a percent. Answers are on page 107.

1.	.45 =	.08 =	.015 =	$.66\frac{2}{3}$ =
2.	.6 =	.15 =	$.08\frac{1}{3}$ =	.96 =
3.	.008 =	.4 =	.01 =	.5 =
4.	1.6 =	.24 =	.375 =	.9 =
5.	2.25 =	5. =	.02 =	$.06\frac{2}{3}$ =

Changing Percents to Decimals

To change a percent to a decimal, move the decimal point two places to the **left** and take off the percent sign.

EXAMPLE: Change 85% to a decimal.

85% = 85 % = .85 — *Step 1.* Move the decimal point two places to the left.

Step 2. Take off the percent sign.

Sometimes you will have to put zeros at the left to get two places.

EXAMPLE: Change 3.9% to a decimal.

3.9% = 03.9% = .039 — *Step 1.* Put a zero to the left of 3.9%.

Step 2. Move the decimal point two places to the left.

Step 3. Take off the percent sign.

EXAMPLE: Change 10% to a decimal.

10% = 10.% = .1 — *Step 1.* Move the decimal point two places to the left.

Step 2. Take off the percent sign.

Notice that we took off the zero at the right in .10. The zero does not change the value of .1.

EXAMPLE: Change $22\frac{2}{9}$% to a decimal.

$22\frac{2}{9}\% = 22\frac{2}{9}\% = .22\frac{2}{9}$ — *Step 1.* The decimal point is not written. It is understood to be at the right of 22. Move the decimal point two places to the left.

Step 2. Take off the percent sign.

Change each percent to a decimal. Answers are on page 108.

1.	15% =	6% =	62.5% =
2.	1% =	8.7% =	$16\frac{2}{3}$% =
3.	40% =	125% =	60% =
4.	0.3% =	5% =	1000% =
5.	$12\frac{1}{2}$% =	200% =	$1\frac{1}{4}$% =
6.	6.3% =	17% =	48.4% =
7.	$18\frac{3}{4}$% =	0.03% =	55% =
8.	475% =	$13\frac{7}{8}$% =	37% =
9.	0.46% =	550% =	80% =
10.	$33\frac{1}{3}$% =	79% =	70.9% =

Changing Fractions to Percents

There are two different ways to change a fraction to a percent. Look at the examples on this page carefully. Then choose the method you like better.

One method for changing a fraction to a percent is to find a fraction of 100%. In other words, multiply the fraction by 100%.

EXAMPLE: Change $\frac{3}{5}$ to a percent.

$\frac{3}{5} \times 100\% =$

$\frac{3}{\cancel{5}_1} \times \frac{\cancel{100}^{20}}{1} = \frac{60}{1} = 60\%$

Multiply $\frac{3}{5}$ by 100%. Remember to put the percent sign in your answer.

EXAMPLE: Change $\frac{2}{7}$ to a percent.

$\frac{2}{7} \times 100\% =$

$\frac{2}{7} \times \frac{100}{1} = \frac{200}{7} = 28\frac{4}{7}\%$

Multiply $\frac{2}{7}$ by 100%.

The other method for changing a fraction to a percent is to change the fraction to a decimal first. Then change the decimal to a percent.

EXAMPLE: Change $\frac{1}{4}$ to a percent.

$\frac{1}{4} = 4\overline{)1.00}$ (quotient .25) *Step 1.* Change $\frac{1}{4}$ to a decimal.

.25 = .25 = 25% *Step 2.* Change .25 to a percent.

Change each fraction to a percent. Answers are on page 108.

1.	$\frac{9}{10} =$	$\frac{4}{5} =$	$\frac{2}{9} =$	$\frac{3}{50} =$
2.	$\frac{4}{25} =$	$\frac{1}{8} =$	$\frac{2}{3} =$	$\frac{1}{2} =$
3.	$\frac{5}{6} =$	$\frac{4}{7} =$	$\frac{7}{100} =$	$\frac{3}{8} =$
4.	$\frac{1}{12} =$	$\frac{11}{20} =$	$\frac{3}{16} =$	$\frac{9}{100} =$

Changing Percents to Fractions

Remember that a percent is a kind of fraction. To change a percent to a fraction, write the digits in the percent as the numerator. Write 100 as the denominator. Then reduce the fraction.

EXAMPLE: Change 65% to a fraction.

$\frac{65}{\ }$ *Step 1.* Write 65 as the numerator.

$\frac{65}{100}$ *Step 2.* Write 100 as the denominator.

$\frac{65}{100} = \frac{13}{20}$ *Step 3.* Reduce the fraction by 5.

When a percent has a decimal in it, first change the percent to a decimal. Then change the decimal to a fraction.

EXAMPLE: Change 4.8% to a fraction.

$4.8\% = 04.8 = .048$ *Step 1.* Change 4.8% to a decimal.

$\frac{048}{1000} = \frac{6}{125}$ *Step 2.* Change .048 to a fraction.

Step 3. Reduce $\frac{48}{1000}$ by 8.

When a percent has a fraction in it, write the digits in the percent as the numerator. Write 100 as the denominator. Then **divide** the numerator by the denominator. This operation is complicated. Study the following example carefully.

EXAMPLE: Change $41\frac{2}{3}\%$ to a fraction.

$\frac{41\frac{2}{3}}{\ }$ *Step 1.* Write $41\frac{2}{3}$ as the numerator.

$\frac{41\frac{2}{3}}{100}$ *Step 2.* Write 100 as the denominator.

Step 3. Remember that the line separating the numerator from the denominator means "divided by." Divide $41\frac{2}{3}$ by 100.

$41\frac{2}{3} \div 100 =$

$\frac{125}{3} \div \frac{100}{1} =$

$\frac{\overset{5}{\cancel{125}}}{3} \times \frac{1}{\underset{4}{\cancel{100}}} = \frac{5}{12}$

Study the next example carefully before you try the problems.

EXAMPLE: Change $6\frac{2}{3}\%$ to a fraction.

$\frac{6\frac{2}{3}}{}$ *Step 1.* Write $6\frac{2}{3}$ as the numerator.

$\frac{6\frac{2}{3}}{100}$ *Step 2.* Write 100 as the denominator.

Step 3. Divide $6\frac{2}{3}$ by 100.

$6\frac{2}{3} \div 100 =$

$6\frac{2}{3} \div \frac{100}{1} =$

$\frac{\overset{1}{\cancel{20}}}{3} \times \frac{1}{\underset{5}{\cancel{100}}} = \frac{1}{15}$

Change each percent to a fraction and reduce. Answers are on page 108.

1.	15% =	1% =	48% =	90% =
2.	22% =	8% =	125% =	4% =
3.	2.8% =	62.5% =	1.5% =	10.2% =
4.	.05% =	11.5% =	87.5% =	1.25% =
5.	$16\frac{2}{3}\%$ =	$37\frac{1}{2}\%$ =	$8\frac{1}{3}\%$ =	$6\frac{1}{4}\%$ =
6.	$88\frac{8}{9}\%$ =	$83\frac{1}{3}\%$ =	$57\frac{1}{7}\%$ =	$4\frac{1}{6}\%$ =

7. 12.5% =	44% =	74% =	85% =
8. 64% =	32.8% =	$18\frac{3}{5}\%$ =	27.2% =
9. 46% =	$8\frac{3}{4}\%$ =	31% =	$41\frac{4}{5}\%$ =
10. 56% =	$66\frac{2}{3}\%$ =	24% =	88% =
11. $13\frac{1}{5}\%$ =	280% =	$36\frac{1}{2}\%$ =	$16\frac{1}{4}\%$ =
12. $25\frac{1}{4}\%$ =	98% =	245% =	.82% =
13. 76% =	1020% =	770% =	$5\frac{3}{4}\%$ =

Common Fractions, Decimals, and Percents

The exercise on this page will help you to learn some of the common fractions, decimals, and percents you will use most often in your work. You will see that sometimes it is easier to change a percent to a decimal, and sometimes it is easier to change a percent to a fraction. In the exercise, find the fraction and decimal that are equal to each percent. Complete the exercise. Then check your answers on page 108. Use the correct answers to fill in the chart on the next page. Study the answers carefully. Memorize the fraction and decimal that each percent is equal to.

1. 50%   25%   75% $12\frac{1}{2}\%$

2. $37\frac{1}{2}\%$ $62\frac{1}{2}\%$ $87\frac{1}{2}\%$ $33\frac{1}{3}\%$

3. $66\frac{2}{3}\%$ 20%   40%   60%

4. 80%   10%   30%   70%

5. 90% $16\frac{2}{3}\%$ $83\frac{1}{3}\%$

Percent	Fraction	Decimal
50%	______	______
25%	______	______
75%	______	______
$12\frac{1}{2}\%$	______	______
$37\frac{1}{2}\%$	______	______
$62\frac{1}{2}\%$	______	______
$87\frac{1}{2}\%$	______	______
$33\frac{1}{3}\%$	______	______
$66\frac{2}{3}\%$	______	______
20%	______	______
40%	______	______
60%	______	______
80%	______	______
10%	______	______
30%	______	______
70%	______	______
90%	______	______
$16\frac{2}{3}\%$	______	______
$83\frac{1}{3}\%$	______	______

Finding a Percent of a Number

When you studied fractions you learned that a fraction immediately followed by the word **of** means to multiply. A percent immediately followed by the word **of** also means to multiply. To multiply by a percent, you must first change the percent to a fraction or a decimal. Then multiply by the fraction or decimal.

EXAMPLE: Find 20% of 45.

Using a fraction:

$20\% = \frac{20}{100} = \frac{1}{5}$ *Step 1.* Change 20% to a fraction.

Step 2. Multiply 45 by $\frac{1}{5}$.

$\frac{1}{5} \times 45 =$

$\frac{1}{\cancel{5}_1} \times \frac{\cancel{45}^9}{1} = \frac{9}{1} = 9$

Using a decimal:

$20\% = 20.\% = .2$ *Step 1.* Change 20% to a decimal.

Step 2. Multiply 45 by .2.

$$\begin{array}{r} 45 \\ \times\ .2 \\ \hline 9.0 = 9 \end{array}$$

Solve each problem using either fractions or percents. Answers are on page 109.

1. 25% of 84 = 40% of 70 = 10% of 96 =

2. 60% of 75 = 75% of 64 = 50% of 48 =

3. 15% of 80 = 24% of 200 = 19% of 300 =

4. 80% of 15 = 5% of 120 = 90% of 80 =

5. 6% of 300 = 20% of 115 = 35% of 40 =

6. 200% of 16 = 30% of 90 = 40% of 150 =

7. 125% of 32 = 1000% of 60 = 500% of 12 =

8. 450% of 86 = 340% of 250 = 275% of 144 =

Sometimes decimals are easier to use than fractions. When the percent itself has a decimal, it is easier to change the percent to a decimal than to a fraction.

EXAMPLE: Find 4.5% of 80.

4.5% = 04.5 % = .045 *Step 1.* Change 4.5% to a decimal.

Step 2. Multiply 80 by .045.

$$\begin{array}{r} 80 \\ \times\ .045 \\ \hline 400 \\ 3\,20 \\ \hline 3.600 \end{array} = 3.6$$

Solve each problem by changing the percent to a decimal.

9. 12.5% of 64 = 8.5% of 700 = 1.2% of 1000 =

10. 10.2% of 400 = 8.9% of 600 = 20.2% of 800 =

11. 1.25% of 500 = 0.5% of 200 = 0.45% of 300 =

12. 10.7% of 2000 = 7.5% of 84 = 32.8% of 375 =

13. 30.6% of 25 = 15.5% of 440 = 0.05% of 1000 =

14. 24.8% of 750 = 35.2% of 2025 = 0.9% of 30 =

Percents with fractions in them are the hardest to use. When percents have fractions in them, change the percent to a simple fraction.

EXAMPLE: Find $16\frac{2}{3}\%$ of 180.

$16\frac{2}{3}\% = \frac{1}{6}$ — *Step 1.* Change $16\frac{2}{3}\%$ to a fraction. (See the chart on page 66.)

$\frac{1}{6} \times 180 =$ — *Step 2.* Multiply 180 by $\frac{1}{6}$.

$\frac{1}{\cancel{6}_1} \times \frac{\cancel{180}^{30}}{1} = \frac{30}{1} = 30$

You will not always know what fraction a percent is equal to. In these cases change the percent to an improper fraction and divide the fraction by 100. Then multiply the new fraction by the other number.

EXAMPLE: Find $42\frac{6}{7}\%$ of 35.

$42\frac{6}{7} = \frac{300}{7}$ — *Step 1.* Change $42\frac{6}{7}$ to an improper fraction.

$\frac{\frac{300}{7}}{100}$ — *Step 2.* Write $\frac{300}{7}$ as the numerator and 100 as the denominator.

$\frac{300}{7} \div 100 =$ — *Step 3.* Divide $\frac{300}{7}$ by 100.

$\frac{300}{7} \div \frac{100}{1} =$ — *Step 4.* Multiply 35 by $\frac{3}{7}$.

$\frac{\cancel{300}^{3}}{7} \times \frac{1}{\cancel{100}_1} = \frac{3}{7}$

$\frac{3}{\cancel{7}_1} \times \frac{\cancel{35}^{5}}{1} = \frac{15}{1} = 15$

There is a shorter way to solve this problem. Change the percent to an improper fraction. Then multiply the improper fraction by the other number in the numerator and by 100 in the denominator. You will see that the answer is the same.

$42\frac{6}{7} = \frac{300}{7}$ — *Step 1.* Change $42\frac{6}{7}$ to an improper fraction.

$\frac{\cancel{300}^{3}}{\cancel{7}_1} \times \frac{\cancel{35}^{5}}{\cancel{100}_1} = \frac{15}{1} = 15$ — *Step 2.* Multiply $\frac{300}{7}$ by $\frac{35}{100}$.

Some of the examples are harder than others. Be sure that you understand them all before you try the next problems.

Solve each problem by changing the percent to a fraction.

15. $33\frac{1}{3}\%$ of 660 = $83\frac{1}{3}\%$ of 180 = $66\frac{2}{3}\%$ of 24 =

16. $16\frac{2}{3}\%$ of 300 = $37\frac{1}{2}\%$ of 320 = $58\frac{1}{3}\%$ of 600 =

17. $22\frac{2}{9}\%$ of 180 = $6\frac{1}{4}\%$ of 480 = $20\frac{5}{6}\%$ of 240 =

18. $\frac{1}{2}\%$ of 800 = $\frac{3}{8}\%$ of 400 = $\frac{2}{3}\%$ of 900 =

19. $\frac{3}{4}\%$ of 2000 = $\frac{5}{6}\%$ of 1200 = $\frac{3}{5}\%$ of 600 =

20. $14\frac{2}{7}\%$ of 84 = $13\frac{1}{3}\%$ of 45 = $8\frac{1}{3}\%$ of 144 =

Finding a Percent of a Number: Applications

These problems give you a chance to apply your skills in finding a percent of a number. When you find a percent of a number, you find a part of that number. The answer will have the same label as the number in the problem. For example, if you find a percent of $25, the answer will be measured in dollars. If you find a percent of 150 pounds, the answer will be measured in pounds.

Give each answer the correct label. Answers are on page 109.

1. George took a math test with 60 problems. He got 80% of the problems right. How many problems did he get right?

2. Mr. and Mrs. Soto make $12,500 a year. They spend 35% of their income on food. How much do the Sotos spend on food in a year?

3. Richard weighed 192 pounds. He went on a diet and lost $12\frac{1}{2}$% of his weight. How many pounds did Richard lose?

4. The auditorium of the Midvale Community Center holds 360 people. 90% of the seats were filled at the last town meeting. How many people sat in the auditorium for the meeting?

5. The sales tax in Jeff's state is 6%. He bought a jacket for $25. How much sales tax did Jeff pay on the jacket?

6. There are 145,000 registered voters in Central County. At the last election 65% of the voters went to vote. How many people voted in Central County in the last election?

7. David and Mary Wiley are buying a house for $32,500. They have to make a down payment of 12%. How much is the down payment for the house?

8. 21% of the budget of Midvale goes for education. Last year's budget for Midvale was $3,500,000. How much did the town spend on education last year?

9. The town of Midvale spent 4% of its budget on health services. Their total budget for the year was $3,500,000. How much did Midvale spend on health services last year?

10. Fred makes $240 a week. His employer takes out 20% of Fred's pay for taxes and social security. How much does Fred's employer take out of Fred's pay each week?

11. Elizabeth owes $200 on her credit card. In a month she has to pay a fee of 1.5% of the amount she owes. Find the amount of the fee on the $200.

12. In 1970 there were 230,000 people living in Central County. In 1980 the number of people in the county was 130% of the 1970 number. How many people lived in Central County in 1980?

13. The town of Midvale plans to build a new sports center. It will cost $2,400,000. So far the town has raised $83\frac{1}{3}\%$ of the money they need for the center. How much money has Midvale raised so far?

More Applications: Rounding Off

The answers to many percent problems need to be "rounded off." Some answers have too many decimal places.

EXAMPLE: Find 8% of $4.49.

8% = 08 % = .08	*Step 1.* Change 8% to a decimal.
$4.49 × .08 $.3592	*Step 2.* Multiply $4.49 by .08.

The answer $.3592 has too many decimal places. Money has only two decimal places. Dimes are in the first, or tenths, place. Pennies are in the second, or hundredths, place. Answers to money problems should have no more than two places.

To round off an answer to the nearest cent, look at the digit in the thousandths (third) decimal place.

If the digit in the thousandths place is 5 or more, add 1 to the cents (hundredths).

If the digit in the thousandths place is less than 5, leave the cents alone.

EXAMPLE: Round off $.3592 to the nearest cent.

Step 1. Look at the digit in the thousandths place.

Step 2. Since 9 is in the thousandths place, add 1 to the cents.

$.3592 to the nearest cent is $.36.

EXAMPLE: Find 11% of $9.65 to the nearest cent.

11% = 11 % = .11	*Step 1.* Change 11% to a decimal.
$9.65 × .11	*Step 2.* Multiply $9.65 by .11.
965 965	*Step 3.* Look at the digit in the thousandths place.
$1.0615	*Step 4.* Since the digit in the thousandths place is 1, leave the cents alone.

$1.0615 to the nearest cent is $1.06.

Round off each answer to the nearest cent. Answers are on page 109.

1.	6% of $5.49 =	8% of $2.55 =	5% of $16.33 =
2.	9% of $8.25 =	12% of $9.89 =	15% of $25.44 =
3.	10% of $2.37 =	2% of $4.52 =	14% of $36.70 =
4.	4% of $8.12 =	13% of $25.50 =	3% of $31.60 =
5.	1.5% of $2.49 =	0.5% of $8.50 =	6.8% of $19 =
6.	4.5% of $33.50 =	0.25% of $89 =	12.5% of $47 =
7.	3.7% of $24.60 =	2.2% of $14.90 =	4.3% of $85 =
8.	8.75% of $14 =	6.25% of $98 =	5.8% of $84.20 =

Finding a Percent of a Number: More Applications

In many percent problems, you will use two operations to find the answer. Sometimes you will first find a percent of a number. Then you will add this new amount to an old amount in the problem.

EXAMPLE: George bought a used car for $1600. He fixed the car and sold it for 25% more than he paid for it. How much did George sell the car for?

$25\% = \frac{25}{100} = \frac{1}{4}$ *Step 1.* Change 25% to a fraction.

$\frac{1}{4} \times \$1600 =$ *Step 2.* Multiply $1600 by $\frac{1}{4}$.

$\frac{1}{\cancel{4}_1} \times \frac{\cancel{1600}^{400}}{1} = \$400$ *Step 3.* Add $400 to $1600.

$$\begin{array}{r} \$1600 \\ +\ \ 400 \\ \hline \$2000 \end{array}$$

Sometimes you will first find a percent of a number. Then you will subtract this new amount from an old amount in the problem.

EXAMPLE: Jane bought a coat on sale. The coat used to cost $49. She bought it for 15% off the old price. How much did Jane pay for the coat?

15% = 15 % = .15 *Step 1.* Change 15% to a decimal.

Step 2. Multiply $49 by .15.

Step 3. Subtract $7.35 from $49.

$$\begin{array}{r} \$49 \\ \times\ \ .15 \\ \hline 2\ 45 \\ 4\ 9\ \ \\ \hline \$7.35 \end{array}$$

$$\begin{array}{r} \$49.00 \\ -\ \ 7.35 \\ \hline \$41.65 \end{array}$$

Read each problem carefully. Round off money answers to the nearest cent. **Answers are on page 109.**

1. In 1979 Manny made $12,350 a year. In 1980 he got a 9% raise. How much did Manny make in 1980?

2. Gordon bought a sweater for $19.85. The sales tax in Gordon's state is 6%. What was the price of the sweater, including the sales tax?

3. Naomi makes $224 a week. Her employer takes out 18% of her pay for taxes and social security. How much does Naomi take home each week?

4. In 1970 the Canadian dollar was worth $.92 in U.S. money. In 1979 the Canadian dollar was worth 8% less than in 1970. What was the value of the Canadian dollar in 1979?

5. Lois took a test with 80 questions. She answered 90% of the questions correctly. How many questions did she get wrong?

6. In 1960 there were 24 women working at the Apex Canning Company. In 1980 there were 250% more women at Apex than in 1960. How many women worked at Apex in 1980?

7. A year ago Frank weighed 220 pounds. He went on a diet and lost 15% of his weight. How much did Frank weigh at the end of his diet?

8. Susan bought a clock radio for $24.49. The sales tax in Susan's state is 8%. How much did she pay for the clock radio, including the sales tax?

9. 840 children go to the Oakdale School. The day of a snowstorm, 25% of the children were absent. How many children went to school on the day of the storm?

10. In the summer it costs about 24¢ a day to run a refrigerator. In the winter it costs $16\frac{2}{3}$% less than in the summer. How much does it cost to run a refrigerator on a winter day?

11. In 1965 the U.S. government spent about $4 billion on highways. In 1975 the government spent 75% more on highways than in 1965. How much did the U.S. spend in 1975 on highways?

12. Ken bought a portable electric saw on sale. The original price was $24.90. Ken bought the saw for 20% off the old price. How much did Ken pay for the saw?

13. Bill and Heather bought a house for $32,500. They made a down payment of 12%. Find the amount they owed on the house after they made the down payment.

14. A year ago Alice paid $.45 for a quart of milk. This year she pays 18% more than last year. How much does Alice pay for a quart of milk this year?

More Applications: Interest

Interest is money someone pays for using someone else's money. A bank pays you interest for using your money in a savings account. You pay a bank interest for using the bank's money on a loan.

To find interest, multiply the principal by the rate and by the time.

The **principal** is the money you borrow or save.
The **rate** is the percent of the interest.
The **time** is the number of years.

EXAMPLE: Find the interest on $400 at 8% annual interest for one year.

$8\% = \frac{8}{100}$ — *Step 1.* Change 8% to a fraction.

$\frac{\cancel{400}^{4}}{1} \times \frac{8}{\cancel{100}_{1}} \times 1 = \$32$ — *Step 2.* Multiply the principal by the rate and by the time.

EXAMPLE: Find the interest on $2000 at 6.5% annual interest for one year.

6.5% = 06 5% = .065 — *Step 1.* Change 6.5% to a decimal.

Step 2. Multiply the principal by the rate and by the time.

```
   $2000
×   .065
--------
  10 000
  120 00
--------
$130.000
```

Note: When the time is one year, you do not have to multiply by 1.

Find the interest for each of the following. Answers are on page 110.

1. $500 at 6% annual interest for one year.

2. $1000 at 12% annual interest for one year.

3. $1200 at $5\frac{1}{2}$% annual interest for one year.

4. \$700 at 4.8% annual interest for one year.

5. \$3000 at 12.5% annual interest for one year.

6. \$650 at 4.5% annual interest for one year.

7. \$10,000 at 15% annual interest for one year.

8. \$900 at 6.2% annual interest for one year.

9. \$350 at 1.5% annual interest for one year.

10. \$800 at $5\frac{3}{4}$% annual interest for one year.

When the time period for interest is not one year, change the time to a fraction of a year.

EXAMPLE: Find the interest on \$500 at 8% annual interest for 9 months.

$8\% = \frac{8}{100}$ — *Step 1.* Change 8% to a fraction.

$\frac{9}{12} = \frac{3}{4}$ — *Step 2.* Change 9 months to a fraction. Write 9 in the numerator and 12 months (one whole year) in the denominator.

$\frac{\overset{5}{\cancel{500}}}{1} \times \frac{\overset{2}{\cancel{8}}}{\underset{1}{\cancel{100}}} \times \frac{3}{\underset{1}{\cancel{4}}} = \$30$ — *Step 3.* Multiply the principal by the rate and by the time.

EXAMPLE: Find the interest on $900 at 15% annual interest for one year and 8 months.

$15\% = \frac{15}{100}$

Step 1. Change 15% to a fraction.

$1\frac{8}{12} = 1\frac{2}{3} = \frac{5}{3}$

Step 2. Change 1 year and 8 months to a mixed number. Write 1 year as a whole number. Write 8 months as a fraction over 12 months.

$\frac{\overset{9}{\cancel{900}}}{1} \times \frac{\overset{5}{\cancel{15}}}{\underset{1}{\cancel{100}}} \times \frac{5}{\underset{1}{\cancel{3}}} = \$225$

Step 3. Multiply the principal by the rate and by the time.

Find the interest for each of the following. Answers are on page 111.

11. $600 at 4% annual interest for 6 months.

12. $2000 at 8% annual interest for 3 months.

13. $1200 at 7% annual interest for 5 months.

14. $250 at 6% annual interest for 1 year and 4 months.

15. $1000 at 9% annual interest for 2 years and 6 months.

16. $1800 at $4\frac{1}{2}$% annual interest for 10 months.

17. $1500 at 12% annual interest for 1 year and 6 months.

Finding What Percent One Number Is of Another

When you studied fractions you learned how to find what part one number is of another. You made a fraction with the part as the numerator and the whole as the denominator. The steps are almost the same for finding what percent one number is of another. Make a fraction with the part as the numerator and the whole as the denominator. Then change the fraction to a percent.

EXAMPLE: 12 is what percent of 48?

$\frac{12}{48}$ — *Step 1.* Make a fraction with the part (12) over the whole (48).

$\frac{12}{48} = \frac{1}{4}$ — *Step 2.* Reduce the fraction by 12.

$\frac{1}{4} \times 100\% =$ — *Step 3.* Change $\frac{1}{4}$ to a percent.

$$\frac{1}{\cancel{4}_1} \times \frac{\cancel{100}^{25}}{1} = \frac{25}{1} = 25\%$$

In these problems you compare a part to a whole. Often you compare a small number to a bigger number. But sometimes you have to compare a big number to a smaller one.

EXAMPLE: 28 is what percent of 24?

$\frac{28}{24}$ — *Step 1.* In this problem you are comparing 28 to a smaller number. Make a fraction with the part (28) over the whole (24).

$\frac{28}{24} = \frac{7}{6}$ — *Step 2.* Reduce the fraction by 4.

$\frac{7}{6} \times 100\% =$ — *Step 3.* Change $\frac{7}{6}$ to a percent.

$$\frac{7}{\cancel{6}_3} \times \frac{\cancel{100}^{50}}{1} = \frac{350}{3} = 116\frac{2}{3}\%$$

Solve each problem. Answers are on page 111.

1. 16 is what % of 64? 18 is what % of 27? 21 is what % of 35?

2. 36 is what % of 48? 64 is what % of 128? 24 is what % of 60?

3.	35 is what % of 50?	18 is what % of 48?	240 is what % of 300?
4.	45 is what % of 54?	10 is what % of 120?	19 is what % of 190?
5.	25 is what % of 80?	14 is what % of 70?	140 is what % of 400?
6.	8 is what % of 120?	135 is what % of 150?	280 is what % of 320?
7.	45 is what % of 500?	25 is what % of 200?	9 is what % of 75%
8.	100 is what % of 80?	40 is what % of 16?	150 is what % of 90?
9.	66 is what % of 48?	165 is what % of 75?	105 is what % of 35?
10.	63 is what % of 70?	8 is what % of 400?	19 is what % of 19?

Percent Applications

These problems give you a chance to apply your skills in finding what percent one number is of another. In these problems compare one number to another by making a fraction. Then change the fraction to a percent. The answers to these problems are all measured in percents. **Answers are on page 111.**

1. The Alonsos make $840 a month. They spend $210 a month for rent. What percent of their income do the Alonsos spend for rent?

2. Last year Silvia weighed 160 pounds. She went on a diet and lost 20 pounds. What percent of her weight did Silvia lose?

3. 350 people work at the Allied Paper Products factory. 280 workers at the factory are men. What percent of the workers at the factory are men?

4. In 1979 Cecilia paid 64¢ for a loaf of bread. In a year the price of a loaf of bread went up 8¢. The increase was what percent of the 1979 price?

5. Jack put $1200 in a savings account for one year. He got $72 in interest for the money. The interest was what percent of the amount Jack put in the bank?

6. In 1970 an ounce of gold was worth $35. In 1979 an ounce of gold was worth $500. The 1970 value was what percent of the 1979 value?

7. Sam took a test with 120 problems. He got 102 problems right. What percent of the problems did Sam get right?

8. John borrowed $2000. He had to pay $240 in interest on the loan. The interest was what percent of the loan?

9. Ten years ago there were 25,000 people living in Midvale. Now there are 10,000 more people living in Midvale. The increase in population is what percent of the population ten years ago?

10. There are 18 students in Mr. Hoffman's night school math class. One evening 6 students were absent. What percent of the students were absent?

11. Naomi works 40 hours a week. She spends about 15 hours every week typing. What percent of her work week does Naomi spend typing?

12. In 1978 the U.S. produced about 48 million tons of wheat. In 1979 the U.S. produced about 56 million tons of wheat. The amount of wheat produced in 1978 was what percent of the amount produced in 1979?

13. George and Martha drove 480 miles to visit their grandchildren. On their way they stopped for lunch after they drove 200 miles. What percent of the total trip had they finished by lunchtime?

More Percent Applications

In some problems you have to compare the difference between two amounts to an original amount. First subtract to find the difference. Then make a fraction with the difference as the numerator and the original amount as the denominator. Change the fraction to a percent.

EXAMPLE: In 1975 Joe made $12,000. In 1980 he made $15,000. By what percent did Joe's income increase from 1975 to 1980?

$$\begin{array}{r} \$15{,}000 \\ -\ 12{,}000 \\ \hline \$\ 3{,}000 \end{array}$$

Step 1. Find how much Joe's income went up. Subtract his old income from his new income.

$$\frac{3{,}000}{12{,}000} = \frac{1}{4}$$

Step 2. Make a fraction with the difference ($3,000) over the original income ($12,000) and reduce.

$$\frac{1}{\cancel{4}_1} \times \frac{\cancel{100}^{25}}{1} = 25\%$$

Step 3. Change $\frac{1}{4}$ to a percent.

EXAMPLE: Bea bought a suitcase on sale for $40. Before the sale the suitcase cost $50. Find the percent of discount on the original price.

$$\begin{array}{r} \$50 \\ -\ 40 \\ \hline \$10 \end{array}$$

Step 1. Find the difference (the discount). Subtract the new price from the old price.

$$\frac{10}{50} = \frac{1}{5}$$

Step 2. Make a fraction with the difference ($10) over the original price ($50) and reduce.

$$\frac{1}{\cancel{5}_1} \times \frac{\cancel{100}^{20}}{1} = 20\%$$

Step 3. Change $\frac{1}{5}$ to a percent.

Study the examples carefully. Then complete the exercise on the next page. Answers are on page 111.

1. In 1970 there were 240,000 people living in Central County. In 1980 there were 330,000 people living there. By what percent did the population increase from 1970 to 1980?

2. Last year Jack made $3.60 an hour. This year he makes $4.20 an hour. By what percent did his wage increase?

3. In 1979 there were 128 fires in Midvale. In 1980 there were only 96 fires. By what percent did the number of fires drop from 1979 to 1980?

4. Don manages a record shop. He pays $3.50 for a long-playing record. He charges customers $4.90 for a record. By what percent does Don mark up the price?

5. There are usually 20 students in Mr. Green's math class. The day of a rainstorm there were only 11 students in the class. What percent of the class was absent?

6. Last year Celeste paid $200 a month for rent. This year she pays $218 a month. By what percent did her rent increase?

7. The Eleventh Street Tenants' Organization started with 18 members. In six months there were 63 members. By what percent did the organization grow in six months?

8. Jim bought a calculator on sale. Before the sale the calculator cost $24. The sale price was $16. Find the percent of discount on the original price.

9. The price of a portable radio is $25.00. Last year the same radio cost $20.00. By what percent has the price of the portable radio increased in the past year?

10. The Midtown Hotel has 45 rooms. 36 of the rooms are being used by travelers. What percent of the rooms are empty?

11. There are 75,000 registered voters in Central County. 3,000 of the registered voters did not vote in the last election. Find the percent of registered voters who voted in Central County's last election.

12. Last year 132 students signed up for night school at the Midvale Community Center. This year only 120 students signed up. By what percent has the number of night school students decreased?

13. The regular price for a kitchen table is $150.00. The local department store is selling the table for $132.00. Find the percent of the discount on the table.

Finding a Number When a Percent of It Is Given

When you studied fractions you learned how to find a whole when a part of the whole was given. We called these problems "backwards multiplication." To find the missing number you divide by the fraction. The steps are almost the same for finding a number when a percent of the number is given. Change the percent to a fraction or a decimal. Then divide the number you have by the fraction or decimal.

EXAMPLE: 30% of what number is 21?

If we had the missing number, 30% multiplied by the number would give us 21. To find the missing number, divide 21 by 30%. (When you go on to study algebra, you will often use this method to solve problems. It is called using opposite operations.)

Using a fraction:

$30\% = \frac{30}{100} = \frac{3}{10}$ *Step 1.* Change 30% to a fraction.

$21 \div \frac{3}{10} =$ *Step 2.* Divide 21 by $\frac{3}{10}$.

$\frac{\overset{7}{\cancel{21}}}{1} \times \frac{10}{\underset{1}{\cancel{3}}} = \frac{70}{1} = 70$

Using a decimal:

$30\% = .30\% = .3$ *Step 1.* Change 30% to a decimal.

$.3\overline{)21.0}$ = 70. *Step 2.* Divide 21 by .3.

To check this problem, find 30% of 70. The answer should be 21.

$30\% = \frac{3}{10}$

$\frac{3}{10} \times 70 = \frac{3}{\underset{1}{\cancel{10}}} \times \frac{\overset{7}{\cancel{70}}}{1} = \frac{21}{1} = 21$

Solve each problem. Answers are on page 112.

1. 80% of what number is 120? $12\frac{1}{2}\%$ of what number is 15?

2. 25% of what number is 19? $16\frac{2}{3}$% of what number is 25?

3. 40% of what number is 34? $37\frac{1}{2}$% of what number is 21?

4. 50% of what number is 165? 60% of what number is 150?

5. $83\frac{1}{3}$% of what number is 45? 75% of what number is 600?

6. 2.5% of what number is 160? $87\frac{1}{2}$% of what number is 140?

7. 62.5% of what number is 15? 4.8% of what number is 240?

8. 100% of what number is 500? 8.5% of what number is 340?

Percent Applications

These problems give you a chance to apply your skills in finding a number when a percent of the number is given. The number you find will be measured in the same units as the number in the problem.

Give each answer the correct label. Answers are on page 112.

1. Joan spends $185 a month for rent. Her rent is 25% of her income. What is Joan's income for a month?

2. 80% of the members of a bus drivers' union voted to strike. 216 members voted to strike. How many members are there in the union?

3. The sales tax in Henry's state is 6%. Henry paid $5.34 in sales tax for a new suit. How much was the suit?

4. Celeste works on a commission rate. She gets 5% of the value of the shoes she sells. Last week she made $97.15 in commissions. What was the value of the shoes she sold last week?

5. Albert got 34 problems right on a math test. He got 85% of the problems on the test right. How many problems were on the test?

6. In a week Mr. and Mrs. Henning spend $99.54 for food for their family. Food is 35% of their budget. How much is the Henning's budget each week?

7. Joe runs a bicycle repair shop. On Monday he fixed 18 bikes. These bikes were 12% of the bikes in the shop. How many bikes were in the shop?

8. Fred had to pay $287.50 interest on a loan. The interest rate on his loan was 11.5%. What was the total amount of the loan?

9. Ten years ago there were 1430 children in school in Midvale. Those children were 65% of the number of children in Midvale schools today. How many children go to Midvale schools today?

10. In an election for the state representative from Central County, Don Johnson got 52% of the votes. He got 26,650 votes. How many people voted in the election?

11. In 1969 the average American ate 10.8 pounds of cheese in a year. That amount was 60% of the 1979 average. How many pounds of cheese did the average American eat in 1979?

12. In 1973 a barrel of crude oil cost $4.50. That was 15% of the 1979 price. How much did a barrel of crude oil cost in 1979?

13. $16\frac{2}{3}$% of the tickets for an all-city basketball tournament were sold the first hour they went on sale. That hour 985 tickets were sold. What was the total number of tickets available for the tournament?

Mixed Percent Problems

You have studied three basic types of percent problems.

Type A, finding a percent of a number. In these problems you have a whole and you want to find a part of it. To find a percent of a number, first change the percent to a fraction or a decimal. Then **multiply** by the fraction or decimal.

Type B, finding what percent one number is of another. In these problems the percent is not given. To find what percent one number is of another, first make a fraction with the part over the whole. Then change the fraction to a percent.

Type C, finding a number when a percent of it is given. In these problems you have the part but not the whole. To find the missing whole, first change the percent to a fraction or a decimal. Then **divide** by the fraction or decimal.

For each problem tell if it is Type A, B, or C. Then solve each problem. Answers are on page 113.

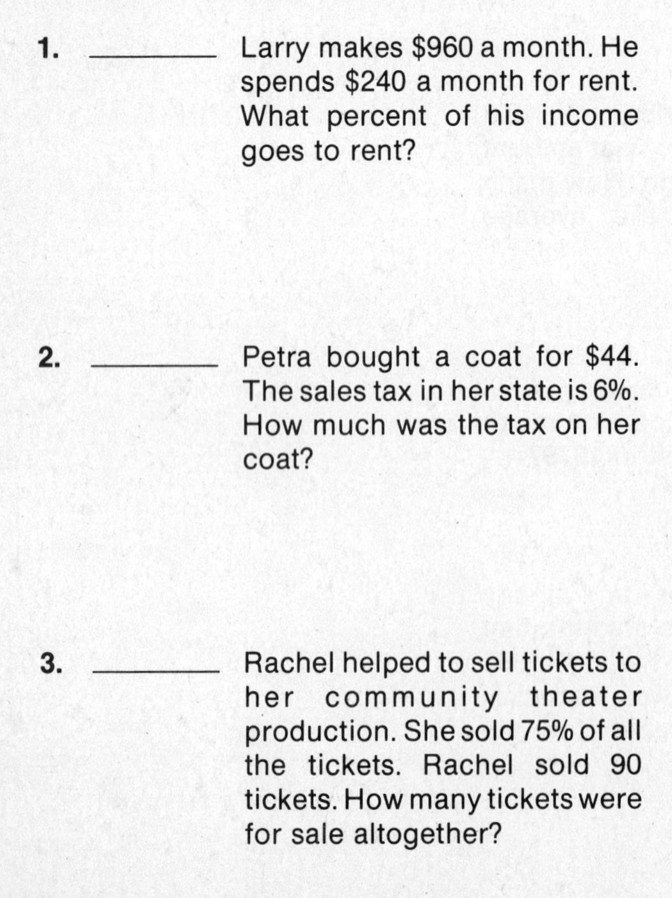

Type

1. ________ Larry makes $960 a month. He spends $240 a month for rent. What percent of his income goes to rent?

2. ________ Petra bought a coat for $44. The sales tax in her state is 6%. How much was the tax on her coat?

3. ________ Rachel helped to sell tickets to her community theater production. She sold 75% of all the tickets. Rachel sold 90 tickets. How many tickets were for sale altogether?

Type

4. ________ Fred took a test with 80 questions. He got 75% of the questions right. How many questions did he get right?

5. ________ There are 20 students in Mr. Wiley's math class. 11 of the students are women. The women make up what percent of the class?

6. ________ Last year Dorothy paid $12 a month for phone service. That amount was 80% of what she pays this year. How much does Dorothy pay each month for her phone this year?

7. ________ The Midvale budget for the year is $3,500,000. 3% of the budget goes for recreation. How much does Midvale spend on recreation in a year?

8. ________ Ann and Tom have paid 60% of their mortgage. So far they have paid $18,000. What was the total amount of their mortgage?

9. ________ Mark started working at a salary of $180 a week. In six months he got a $9 a week raise. The raise was what percent of his old salary?

10. ________ There are 650 employees at Pulp Paper Products. 84% of the employees belong to the union. How many employees belong to the union?

Percent Review

These problems will tell you which parts of the percent section of this book you need to review. When you finish, check the chart to see which pages you need to review.

1. Change each decimal to a percent.

$.65 =$ $\quad$ $.02 =$ $\quad$ $.475 =$ $\quad$ $.09\frac{1}{3} =$

2. Change each percent to a decimal.

$92\% =$ $\quad$ $4\% =$ $\quad$ $0.9\% =$ $\quad$ $16\frac{2}{3}\% =$

3. Change each fraction to a percent.

$\frac{12}{25} =$ $\quad$ $\frac{5}{12} =$ $\quad$ $\frac{5}{8} =$ $\quad$ $\frac{2}{13} =$

4. Change each percent to a fraction.

$36\% =$ $\quad$ $8.5\% =$

$28\frac{4}{7}\% =$ $\quad$ $5\frac{5}{9}\% =$

5. Find 40% of 65.

6. Find 35% of 200.

7. Find $16\frac{2}{3}\%$ of 120.

8. Find $87\frac{1}{2}\%$ of 160.

9. Find 125% of 48.

10. Find 350% of 80.

11. Find 6.4% of 300.

12. Find 2.8% of 250.

13. Janet took a test with 40 questions. She got 90% of the questions right. How many questions did she get right?

14. George and Margaret are buying a house for $35,000. They have to make a down payment of 11%. How much is the down payment for the house?

15. Find 8% of $6.29. Round off to the nearest cent.

16. Find 2.5% of $3.16. Round off to the nearest cent.

17. Virginia bought a T.V. on sale. The original price was $145. The T.V. was on sale for 15% off the old price. How much did Virginia pay for the T.V.?

18. In 1970 the town of Midvale spent $3,000,000 on education. In 1980 they spent 40% more than in 1970. How much did they spend on education in 1980?

19. Find the interest on $600 at 9% annual interest for one year.

20. Find the interest on $3000 at 11% annual interest for eight months.

21. 27 is what % of 72?

22. 18 is what % of 36?

23. 16 is what % of 80?

24. 35 is what % of 42?

25. Jim makes $250 a week. His employer takes out $35 for taxes and social security. What percent of Jim's pay goes to taxes and social security?

26. Ann put $800 in a savings account for a year. She got $48 in interest. The interest was what percent of the amount Ann put in the bank?

27. Usually there are 25 people at the meetings of the Eleventh Street Tenants' Organization. At the last meeting there were only 16 people. What percent of the usual number was absent?

28. Last year Sue paid $160 a month for rent. This year she pays $184 a month. By what percent did her rent go up?

29. 25% of what number is 12?

30. 60% of what number is 90?

31. 87.5% of what number is 35?

32. $83\frac{1}{3}\%$ of what number is 75?

33. Jeff works on a commission rate. He gets 5% of the value of the clothes he sells. Last week he made $125.30 in commissions. What was the value of the clothes he sold last week?

34. Petra got 80% of the problems right on a test. She got 52 problems right. How many problems were on the test?

Check your answers on page 113. Circle the problems you missed on the chart below. Review the pages that show how to work the problems you missed. Then try the problems again.

Problem Number	Review Pages	Problem Number	Review Pages
1	58	18	75 - 77
2	59	19	78 - 80
3	61	20	78 - 80
4	62 - 64	21	81 - 82
5	67 - 70	22	81 - 82
6	67 - 70	23	81 - 82
7	67 - 70	24	81 - 82
8	67 - 70	25	83 - 84
9	67 - 70	26	83 - 84
10	67 - 70	27	85 - 87
11	67 - 70	28	85 - 87
12	67 - 70	29	88 - 89
13	71 - 72	30	88 - 89
14	71 - 72	31	88 - 89
15	73 - 74	32	88 - 89
16	73 - 74	33	90 - 91
17	75 - 77	34	90 - 91

Answers

Answers

pages 2-6

1. 76.3 8.067 1.2804

2. **a.** three four one
b. two one four

3. 9.007 .0038 14.0601 80.0503

4. **a.** four tenths one and twenty-seven hundredths
b. eighteen hundredths two and fifteen thousandths
c. four ten-thousandths sixteen and nine hundredths
d. twelve millionths one hundred twenty and fifty-six ten-thousandths

5. **a.** .16
b. 10.5
c. .00022
d. 309.008
e. .000408

6. **a.** .8 .061 .203
b. .2 .41 .092

7. **a.** .3 .808 .11
b. .71 .06 .401

8. **a.** $\frac{2}{25}$ $\frac{3}{10}$ $4\frac{1}{5}$ $9\frac{1}{8}$
b. $\frac{9}{250}$ $8\frac{9}{2000}$ $\frac{3}{500}$ $12\frac{2}{5}$

9. **a.** .6 .7 .24 $.22\frac{2}{9}$
b. .15 $.37\frac{1}{2}$ or .375 $.06\frac{2}{3}$ $.41\frac{2}{3}$

10. 1.5 **11.** 1.352
12. 1.428 **13.** .9521
14. 12.856 **15.** 29.59
16. 15.389 **17.** 4.243
18. 20 million
19. 23.5 gallons
20. .02 **21.** .56
22. 7.825 **23.** 5.72
24. 13.291 **25.** 2.16
26. 4.595 kilograms
27. 1.97 billion
28. 3.656 **29.** 15.68
30. 1.675 **31.** .0288
32. .117 **33.** 3.2364
34. 3.2 **35.** 670 **36.** 3450
37. $5.72
38. 164.64 miles
39. .53 **40.** 8.6
41. 4.2 **42.** 35
43. 610 **44.** .23
45. .08 **46.** 120
47. 40 **48.** 45
49. .048 **50.** .062 **51.** .004357
52. $3.80 **53.** 85 kilograms

page 9

1. 904.6 7.3
   40.9 6.622 86.33 3.8
2. two three five
   two one two
   one two four
   three four none
3. 90.085 96.3 5.9 200.04
   4.306 30.0105 2,060.7 750.8
   2.358 6 .00705 60.2

page 10

1. two tenths six tenths
2. three hundredths forty-nine hundredths
3. eighteen thousandths two hundred seven thousandths
4. six thousandths four hundred eighty-three thousandths
5. fifty-eight ten-thousandths one hundred thirty-two ten-thousandths
6. three hundred twenty-five hundred-thousandths five millionths
7. nine hundred-thousandths one and six tenths
8. two hundred eight and four tenths thirteen and nine hundredths

page 11

1. .2 7.05
2. .015 40.3
3. .06 13.007
4. .0019 .000009
5. 2.000809 **6.** 512.4
7. 800.0006 **8.** .00018

page 12

1. .75 .2 .08
2. .3 .05 .072
3. .53 .22 .453
4. .3 .61 .1024
5. .4 .21 .403
6. .201 .57 .1101
7. .71 .606 .44
8. .5205 .8 .3

page 13

1. $\frac{1}{25}$ $\frac{1}{2}$ $7\frac{9}{25}$ $15\frac{5}{8}$
2. $\frac{1}{8}$ $\frac{1}{125}$ $1\frac{1}{2500}$ $28\frac{7}{50}$
3. $\frac{7}{2000}$ $\frac{17}{40}$ $3\frac{17}{20}$ $16\frac{33}{50}$
4. $\frac{16}{25}$ $\frac{3}{4}$ $9\frac{4}{25}$ $34\frac{1}{500}$
5. $\frac{13}{200}$ $\frac{3}{4000}$ $10\frac{3}{125}$ $8\frac{3}{50,000}$

page 15

1. .75 .4 .3 $.33\frac{1}{3}$
2. $.16\frac{2}{3}$ .35 .5 $.62\frac{1}{2}$ or .625
3. .32 $.55\frac{5}{9}$ .9 $.08\frac{1}{3}$

4. $.28\frac{4}{7}$ $.12\frac{1}{2}$ or .125 .68 $.66\frac{2}{3}$

5. .8 $.31\frac{1}{4}$ $.87\frac{1}{2}$ or .875 .05

6. .65 $.81\frac{1}{4}$ or .8125 $.59\frac{1}{7}$ $.77\frac{7}{9}$

7. 1.25 $.37\frac{1}{2}$ or .375 .36 .14

8. .7 $1.66\frac{2}{3}$ .95 .2

9. .46 .96 1.55 $.44\frac{4}{9}$

10. .1 .62 .15 3.3

pages 16-17

1. 1.178 2.388 **2.** 2 1.12
3. 2.074 1.3116 **4.** 1.262 .8855
5. .8998 .6456 **6.** .989 .36
7. 61.06 8.104 **8.** 95.504 9.356
9. 5.057 13.97 **10.** 13.346 31.444
11. 81.09 3.364 **12.** 4.4875 7.8793
13. 139.084 21.169

pages 18-19

1. 11.3 pounds **2.** 135.3 million
3. 29,591.3 miles **4.** 6.7 million
5. 65.2°
6. 21.6 million **7.** 9.785 kilograms
8. 32.9 miles **9.** $13.1 billion
10. 16.75 hours **11.** $508.75

page 20

1. 4.752 9.85 .237
2. .403 2.932 .007
3. .664 4.434 .335
4. 5.02 1.936 7.76
5. .017 19.1 .162
6. .016 .6593 11.7
7. 7.08 .031 4.495

pages 21-22

1. 5.3 million **2.** $.75 million
3. 55.8 people **4.** .45 meters
5. 15.3 pounds **6.** .002
7. 12.8 tons **8.** $50.60
9. 3 million barrels **10.** 5.8 minutes
11. 70.6 million **12.** $64.15

pages 23-24

1.
$$\begin{array}{r} 4.16 \\ +\ 5.09 \\ \hline \end{array}$$
9.25 billion barrels

2.
$$\begin{array}{r} 6.09 \\ +\ .39 \\ \hline \end{array}$$
6.48 billion barrels

3.
```
  9.25
- 6.48
------
```
2.77 billion barrels

4.
```
  4.16
- 3.85
------
```
.31 billion barrels

5.
```
  8.19
- 5.09
------
```
3.1 billion barrels

6.
```
  7.68
- 3.85
------
```
3.83 billion barrels

7.
```
  8.19
-  .68
------
```
7.51 billion barrels

8.
```
   3.85
+  8.19
-------
```
12.04 billion barrels

9.
```
  7.68
+  .68
------
```
8.36 billion barrels

10.
```
  12.04
-  8.36
-------
```
3.68 billion barrels

11.
```
  7.68
- 6.09
------
```
1.59 billion barrels

12.
```
  .68
- .39
-----
```
.29 billion barrels

13.
```
  6.09
- 4.16
------
```
1.93 billion barrels

14.
```
  5.09
-  .39
------
```
4.7 billion barrels

pages 26-27

1. 10.8 5.84 .36
2. .056 3.68 .18
3. 4.272 .6904 22.62
4. 8.76 18.2 5.742
5. .0152 .00045 .00192
6. 431.2 4.095 9.72
7. .0065 .1599 51.264

8. 21.3 430.1 62.82
9. 35.968 1.7227 11.232
10. 4.755 73.346 2.875
11. .00074 .00144 .0864
12. 95.2 21.63 5400
13. .02072 64.96 .037

page 28

1. 4 17 2.8 14.8
2. 9.25 667 .8 .37
3. 62 453 12.9 570
4. 402 58 34.4 1660
5. 1485 360 23,500 705
6. 2539 85 3600 42,120

pages 29-30

1. $215.60 **2.** 213.6 miles
3. 58.5 kilograms **4.** 10.4 kilometers
5. 68.5 pounds **6.** $5.75
7. 167.64 centimeters **8.** $89.25
9. $4.95 **10.** 247 miles
11. 3.24 kilograms **12.** $1
13. $25.65 **14.** 60.8 inches

pages 31-32

1. 1.6 × 40.7 = **65.12 miles**
1.41 × 41.3 = **58.233 miles**
1.05 × 41.0 = **43.05 miles**
.92 × 39.2 = **36.064 miles**

2.
$$\begin{array}{r} 65.120 \\ -\ 36.064 \\ \hline \end{array}$$
29.056 miles

3. 1.6 × 65.12 = **104.192 kilometers**
1.6 × 58.233 = **93.1728 kilometers**
1.6 × 43.05 = **68.88 kilometers**
1.6 × 36.064 = **57.7024 kilometers**

4. .33 × 65.12 = **21.4896 billion miles**
.30 × 58.233 = **17.4699 billion miles**
.28 × 43.05 = **12.054 billion miles**
.27 × 36.064 = **9.73728 billion miles**

5.
$$\begin{array}{r} 17.4699 \\ -\ 12.0540 \\ \hline \end{array}$$
5.4159 billion miles

6. **330 million riders**
300 million riders
280 million riders
270 million riders

7.
$$\begin{array}{r} 330 \\ -\ 270 \\ \hline \end{array}$$
60 million riders

8. 280
− 270
10 million riders

9. $.06 × 36 = **$2.16**

pages 33-34

1. 4.8 .63 .029
2. 2.38 .843 .57
3. .014 3.9 .028
4. 30.2 .082 .63
5. 124.6 30.09 21.03
6. .903 .029 2.16
7. .0004 43.95 .38
8. .861 26.001 .472
9. 1.4 .73 .4762
10. .81 7.1 57.31
11. 2.1 .023 1.3
12. .41 37.3 86.8

pages 36-37

1. 4.6 .32 6.7
2. .59 9.4 33
3. 63 4.9 .96
4. .58 8.5 470
5. .43 3.9 67
6. 35 47 58
7. 5.3 2610 13.4
8. 62 83 .93
9. .43 6.4 7.5
10. 260 5420 385
11. 496 7600 490
12. .06 5 80
13. .013 .42 70

pages 38-39

1. 15 80 825
2. 12 400 3000
3. 250 175 400
4. 80 240 165
5. 7100 270 28,000
6. 63,750 80 370
7. 12 410 130
8. 1500 200 370
9. 1460 140 270
10. 850 56,000 240

page 40

1. .43 1.8 .06 .0034
2. .328 .9 1.78 .015
3. .192 .0187 .21 .08
4. .0047 .065 .00029 3.64
5. .0284 .00439 .35 .0026
6. 3.98 .0009 .04216 .9235

pages 42-43

1. $1.60 **2.** $.62
3. .79 kilograms **4.** .42 meters
5. 13.4 pounds **6.** $.65 million
7. 60 kilograms **8.** 75 kilometers
9. 40 gallons **10.** 23.4 miles
11. 38.5 hours **12.** 8.848 kilometers

pages 44-45

1. 11.5 miles per gallon
24.6 miles per gallon
15.9 miles per gallon
21.2 miles per gallon
2. Dave
3. Pete
4. Mark
5. John
6.
$$\begin{array}{r} 24.6 \\ -\ 11.5 \\ \hline \end{array}$$
13.1 miles per gallon

7. $1.05
$1.10
$1.08
$1.12
8. Dave
9. John
10. 2656.9 miles
11. 133.2 gallons
12. $146.11
13.
$$\begin{array}{r} \$61.04 \\ -\ 12.96 \\ \hline \$48.08 \end{array}$$

pages 46-50

1. 5.18 4.7
2. **a.** one three two
b. four two three
3. 60.04 8.005 300.29 7.043
4. **a.** eight tenths nine and thirty-five hundredths
b. fifty-six thousandths twelve and one hundredth
c. one hundred three ten-thousandths thirty and two hundred twenty-six thousandths
d. three hundred twenty-five hundred thousandths two and nine millionths
5. **a.** 8.14 **b.** .0039
c. 105.007 **d.** .04708
e. 72.06
6. **a.** .4 .018 .7
b. .63 .50 .14
7. **a.** .76 .22 .46
b. .051 .17 .32
8. **a.** $\frac{7}{10}$ $4\frac{1}{4}$ $6\frac{6}{25}$ $\frac{1}{20,000}$
b. $\frac{3}{625}$ $\frac{7}{8}$ $\frac{9}{500}$ $12\frac{8}{25}$

9. **a.** .35 $.42\frac{6}{7}$ $.91\frac{2}{3}$ .16

b. $.31\frac{1}{4}$ $.11\frac{1}{9}$ .18 $.12\frac{1}{2}$ or .125

10. 1.7 **11.** 2.086
12. .4079 **13.** 1.027
14. 20.436 **15.** 22.1
16. 15.875 **17.** 31.926
18. 67.7¢ **19.** 7.25 meters
20. .504 **21.** 8.74
22. 1.018 **23.** 7.84
24. 3.963 **25.** 5.13
26. 5.6 miles
27. $.45 million
28. 2.934 **29.** 41.85
30. 2.856 **31.** .0558
32. .0481 **33.** 1.524
34. 28 **35.** 80 **36.** 468.1
37. 72 kilograms **38.** $46.50
39. 4.9 **40.** .062
41. 3.8 **42.** 16
43. 520 **44.** .42
45. .09 **46.** 230
47. 400 **48.** 120
49. .62 **50.** .17 **51.** .0032
52. $.70 **53.** .73 meter

pages 52-55

1. 40% 4% 22.5% $6\frac{1}{4}$%

2. .3 .09 $.03\frac{1}{3}$ 2.5

3. 65% $87\frac{1}{2}$% or 87.5% $43\frac{3}{4}$% 18%

4. $\frac{4}{25}$ $\frac{9}{200}$ $\frac{4}{9}$ $\frac{7}{24}$

5. 21 **6.** 31 **7.** 24 **8.** 250
9. 27 **10.** 165 **11.** 4.6 **12.** 8.4
13. $4200 **14.** 253,000 people
15. $.45 **16.** $.17 **17.** $21.06
18. $174.30 **19.** $48 **20.** $65
21. 40% **22.** $66\frac{2}{3}$%
23. 90% **24.** 32%
25. 10% **26.** 60%
27. 75% **28.** 15%
29. 135 **30.** 96
31. 120 **32.** 140
33. 168 members **34.** $700

pages 58

1. 45% 8% 1.5% $66\frac{2}{3}$%

2. 60% 15% $8\frac{1}{3}$% 96%

3. .8% 40% 1% 50%
4. 160% 24% 37.5% 90%

5. 225% 500% 2% $6\frac{2}{3}$%

page 60

1. .15 .06 .625

2. .01 .087 $.16\frac{2}{3}$

3. .4 1.25 .6

4. .003 .05 10

5. $.12\frac{1}{2}$ 2 $.01\frac{1}{4}$

6. .063 .17 .484

7. $.18\frac{3}{4}$ .0003 .55

8. 4.75 $.13\frac{7}{8}$ .37

9. .0046 5.5 .8

10. $.33\frac{1}{3}$ .79 .709

page 61

1. 90% 80% $22\frac{2}{9}$ 6%

2. 16% $12\frac{1}{2}\%$ or 12.5% $66\frac{2}{3}\%$ 50%

3. $83\frac{1}{3}\%$ $57\frac{1}{7}\%$ 7% $37\frac{1}{2}\%$ or 37.5%

4. $8\frac{1}{3}\%$ 55% $18\frac{3}{4}\%$ 9%

pages 63-64

1. $\frac{3}{20}$ $\frac{1}{100}$ $\frac{12}{25}$ $\frac{9}{10}$

2. $\frac{11}{50}$ $\frac{2}{25}$ $1\frac{1}{4}$ $\frac{1}{25}$

3. $\frac{7}{250}$ $\frac{5}{8}$ $\frac{3}{200}$ $\frac{51}{500}$

3. $\frac{1}{2000}$ $\frac{23}{200}$ $\frac{7}{8}$ $\frac{1}{80}$

5. $\frac{1}{6}$ $\frac{3}{8}$ $\frac{1}{12}$ $\frac{1}{16}$

6. $\frac{8}{9}$ $\frac{5}{6}$ $\frac{4}{7}$ $\frac{1}{24}$

7. $\frac{1}{8}$ $\frac{11}{25}$ $\frac{37}{50}$ $\frac{17}{20}$

8. $\frac{16}{25}$ $\frac{41}{125}$ $\frac{93}{500}$ $\frac{34}{125}$

9. $\frac{23}{50}$ $\frac{7}{80}$ $\frac{31}{100}$ $\frac{209}{500}$

10. $\frac{14}{25}$ $\frac{2}{3}$ $\frac{6}{25}$ $\frac{22}{25}$

11. $\frac{33}{250}$ $2\frac{4}{5}$ $\frac{73}{200}$ $\frac{13}{80}$

12. $\frac{101}{400}$ $\frac{49}{50}$ $2\frac{9}{20}$ $\frac{41}{5000}$

13. $\frac{19}{25}$ $10\frac{2}{5}$ $7\frac{7}{10}$ $\frac{1}{16}$

page 65

1. .5, $\frac{1}{2}$ .25, $\frac{1}{4}$ .75, $\frac{3}{4}$ $.12\frac{1}{2}$ or .125, $\frac{1}{8}$

2. $.37\frac{1}{2}$ or 375, $\frac{3}{8}$ $.62\frac{1}{2}$ or .625, $\frac{5}{8}$ $.87\frac{1}{2}$ or .875, $\frac{7}{8}$ $.33\frac{1}{3}$, $\frac{1}{3}$

3. $.66\frac{2}{3}$, $\frac{2}{3}$ .2, $\frac{1}{5}$ .4, $\frac{2}{5}$ .6, $\frac{3}{5}$

4. .8, $\frac{4}{5}$ .1, $\frac{1}{10}$ .3, $\frac{3}{10}$ .7, $\frac{7}{10}$

5. .9, $\frac{9}{10}$ $.16\frac{2}{3}$, $\frac{1}{6}$ $.83\frac{1}{3}$, $\frac{5}{6}$

pages 67-70

1. 21 28 9.6 or $9\frac{3}{5}$
2. 45 48 24
3. 12 48 24
4. 12 6 72
5. 18 23 14
6. 32 27 60
7. 40 600 60
8. 387 850 396
9. 8 59.5 12
10. 40.8 53.4 161.6
11. 6.25 1 1.35
12. 214 6.3 123
13. 7.65 68.2 .5
14. 186 712.8 .27
15. 220 150 16
16. 50 120 350
17. 40 30 50
18. 4 $1\frac{1}{2}$ 6
19. 15 10 $3\frac{3}{5}$
20. 12 6 12

pages 71-72

1. 48 problems **2.** $4375
3. 24 pounds **4.** 324 people
5. $1.50 **6.** 94,250 people
7. $3900 **8.** $735,000
9. $140,000 **10.**$48
11. $3 **12.** 299,000 people
13. $2,000,000

page 74

1. $.33 $.20 $.82
2. $.74 $1.19 $3.82
3. $.24 $.09 $5.14
4. $.32 $3.32 $.95
5. $.04 $.04 $1.29
6. $1.51 $.22 $5.88
7. $.91 $.33 $3.66
8. $1.23 $6.13 $4.88

pages 76-77

1. .09 × $12,350 = $1111.50

```
  $12,350
+   1,111.50
-----------
  $13,461.50
```

(the total $13,461.50 is printed in bold)

2. $.06 \times \$19.85 = \$1.191$

$$\begin{array}{r} \$19.85 \\ +\quad 1.19 \\ \hline \mathbf{\$21.04} \end{array}$$

3. $.18 \times \$224 = \$40.32$

$$\begin{array}{r} \$224.00 \\ -\quad 40.32 \\ \hline \mathbf{\$183.68} \end{array}$$

4. $.08 \times \$.92 = \$.0736$

$$\begin{array}{r} \$.92 \\ -\quad .07 \\ \hline \mathbf{\$.85} \end{array}$$

5. $.9 \times 80 = 72$

$$\begin{array}{r} 80 \\ -\ 72 \\ \hline \end{array}$$

8 questions

6. $2.5 \times 24 = 60$

$$\begin{array}{r} 24 \\ +\ 60 \\ \hline \end{array}$$

84 women

7. $.15 \times 220 = 33$

$$\begin{array}{r} 220 \\ -\quad 33 \\ \hline \end{array}$$

187 pounds

8. $.08 \times \$24.49 = \$1.9592$

$$\begin{array}{r} \$24.49 \\ +\quad 1.96 \\ \hline \mathbf{\$26.45} \end{array}$$

9. $\frac{1}{4} \times 840 = 210$

$$\begin{array}{r} 840 \\ -\ 210 \\ \hline \end{array}$$

630 children

10. $\frac{1}{6} \times 24¢ = 4¢$

$$\begin{array}{r} 24¢ \\ -\quad 4¢ \\ \hline \mathbf{20¢} \end{array}$$

11. $\frac{3}{4} \times 4 = \$3$ billion

$$\begin{array}{r} \$4 \text{ billion} \\ +\quad 3 \text{ billion} \\ \hline \end{array}$$

$7 billion

12. $.2 \times \$24.90 = \$4.98$

$$\begin{array}{r} \$24.90 \\ -\quad 4.98 \\ \hline \mathbf{\$19.92} \end{array}$$

13. $.12 \times \$32{,}500 = \$3{,}900$

$$\begin{array}{r} \$32{,}500 \\ -\quad 3{,}900 \\ \hline \mathbf{\$28{,}600} \end{array}$$

14. $.18 \times \$.45 = \$.081$

$$\begin{array}{r} \$.45 \\ +\quad .08 \\ \hline \mathbf{\$.53} \end{array}$$

pages 78-80

1. $30 **2.** $120 **3.** $66
4. $33.60 **5.** $375
6. $29.25 **7.** $1500

8. \$55.80 **9.** \$5.25
10. \$46 **11.** \$12
12. \$40 **13.** \$35
14. \$20 **15.** \$225
16. \$67.50 **17.** \$270

pages 81-82

1.	25%	$66\frac{2}{3}\%$	60%
2.	75%	50%	40%
3.	70%	$37\frac{1}{2}\%$	80%
4.	$83\frac{1}{3}\%$	$8\frac{1}{3}\%$	10%
5.	$31\frac{1}{4}\%$	20%	35%
6.	$6\frac{2}{3}\%$	90%	$87\frac{1}{2}\%$
7.	9%	$12\frac{1}{2}\%$	12%
8.	125%	250%	$166\frac{2}{3}\%$
9.	$137\frac{1}{2}\%$	220%	300%
10.	90%	2%	100%

pages 83-84

1. 25% **2.** $12\frac{1}{2}\%$
3. 80% **4.** $12\frac{1}{2}\%$
5. 6% **6.** 7%
7. 85% **8.** 12%
9. 40% **10.** $33\frac{1}{3}\%$
11. $37\frac{1}{2}\%$ **12.** $85\frac{5}{7}\%$
13. $41\frac{2}{3}\%$

pages 86-87

1. $\begin{array}{r} 330{,}000 \\ -\ 240{,}000 \\ \hline 90{,}000 \end{array}$ $\frac{90{,}000}{240{,}000} = \frac{3}{8} = \mathbf{37\frac{1}{2}\%}$

2. $\begin{array}{r} \$4.20 \\ -\ 3.60 \\ \hline \$\ .60 \end{array}$ $\frac{.60}{3.60} = \frac{1}{6} = \mathbf{16\frac{2}{3}\%}$

3. $\begin{array}{r} 128 \\ -\ 96 \\ \hline 32 \end{array}$ $\frac{32}{128} = \frac{1}{4} = \mathbf{25\%}$

4. $\begin{array}{r} \$4.90 \\ -\ 3.50 \\ \hline \$1.40 \end{array}$ $\frac{1.40}{3.50} = \frac{2}{5} = \mathbf{40\%}$

5. $\begin{array}{r} 20 \\ -\ 11 \\ \hline 9 \end{array}$ $\frac{9}{20} = \mathbf{45\%}$

6. $\begin{array}{r} \$218 \\ -\ 200 \\ \hline \$\ 18 \end{array}$ $\frac{18}{200} = \mathbf{9\%}$

7. $\begin{array}{r} 63 \\ -\ 18 \\ \hline 45 \end{array}$ $\frac{45}{18} = \frac{5}{2} = \mathbf{250\%}$

8. $\begin{array}{r} \$24 \\ -\ 16 \\ \hline \$\ 8 \end{array}$ $\frac{8}{24} = \frac{1}{3} = \mathbf{33\frac{1}{3}\%}$

9. $\begin{array}{r} \$25 \\ -\ 20 \\ \hline \$\ 5 \end{array}$ $\frac{5}{20} = \frac{1}{4} = \mathbf{25\%}$

10. $\begin{array}{r} 45 \\ -\ 36 \\ \hline 9 \end{array}$ $\frac{9}{45} = \frac{1}{5} = \mathbf{20\%}$

11. $\begin{array}{r} 75{,}000 \\ -\ 3{,}000 \\ \hline 72{,}000 \end{array}$ $\frac{72{,}000}{75{,}000} = \frac{24}{25} = \mathbf{96\%}$

12. $\begin{array}{r} 132 \\ -\ 120 \\ \hline 12 \end{array}$ $\frac{12}{132} = \frac{1}{11} = \mathbf{9\frac{1}{11}\%}$

13. $\begin{array}{r} 150 \\ -\ 132 \\ \hline 18 \end{array}$ $\frac{18}{150} = \frac{3}{25} = \mathbf{12\%}$

pages 88-89

1. 150 120
2. 76 150
3. 85 56
4. 330 250
5. 54 800
6. 6400 160
7. 24 5000
8. 500 4000

pages 90-91

1. $740 **2.** 270 members
3. $89 **4.** $1943
5. 40 problems **6.** $284.40
7. 150 bikes **8.** $2500
9. 2200 children **10.** 51,250 people
11. 18 pounds **12.** $30
13. 5910 tickets

pages 92-93

1. B 25%
2. A $2.64
3. C 120 tickets
4. A 60 questions
5. B 55%
6. C $15
7. A $105,000
8. C $30,000
9. B 5%
10. A 546 employees

pages 94-97

1. 65% 2% 47.5% $9\frac{1}{3}$%

2. .92 .04 .009 $.16\frac{2}{3}$

3. 48% $41\frac{2}{3}$% $62\frac{1}{2}$% or 62.5% $15\frac{5}{13}$%

4. $\frac{9}{25}$ $\frac{17}{200}$ $\frac{2}{7}$ $\frac{1}{18}$

5. 26 **6.** 70 **7.** 20 **8.** 140

9. 60 **10.** 280 **11.** 19.2 **12.** 7

13. 36 questions **14.** $3850

15. $.50 **16.** $.08

17. $123.25

18. $4,200,000

19. $54 **20.** $220

21. $37\frac{1}{2}$% **22.** 50%

23. 20% **24.** $83\frac{1}{3}$%

25. 14% **26.** 6%

27. 36% **28.** 15%

29. 48 **30.** 150

31. 40 **32.** 90

33. $2506 **34.** 65 problems